GIS in Schools

Richard Audet and Gail Ludwig

ESRI PRESS

Environmental Systems Research Institute, Inc.
    GIS in Schools
    ISBN 1-879102-85-4

First printing October 2000

10 9 8 7 6 5 4 3 2 1

Printed in the United States of America.

Published by Environmental Systems Research Institute, Inc., 380 New York Street, Redlands, California 92373-8100.

ESRI Press books are available to resellers worldwide through Independent Publishers Group (IPG). For information on volume discounts, or to place an order, call IPG at 800-888-4741 in the United States, or at 312-337-0747 outside the United States.

Contents

The decade of the 1990s saw a remarkable evolution. In the early nineties, geographic information systems (GIS) were the exclusive domain of high-tech practitioners in business, government, and research. With the explosion of computers and computing power, GIS technology moved into the mainstream of American society. But one province is still underrepresented: the world of K–12 education.

The hardware necessary for GIS was out of reach of K–12 schools in the early nineties. But now, powerful computers abound, Internet access is commonplace, and tech-savvy kids populate classrooms across the land.

In 1992, ESRI began its program to support K–12 education. From the beginning, the message has been: If using powerful tools to work with spatial data can help adults explore information and solve problems by integrating geography with science, math, and communication, it can work for kids as well.

With hardware no longer the ultimate barrier, the controlling factor today is "educational vision." The stories in this volume show that students can do very impressive work, if given the opportunity. Under the guidance of teachers willing to introduce the technology and let them go to work, students in high school, middle school, and even elementary school can wrestle with confusing global data and even address fundamental issues facing the local community. The varied settings, diverse projects, and unique personalities in these chapters show that "any school can use GIS."

In an era when teachers take a public beating all too often, it is heartening that some teachers still model the lifelong learner, the thinking explorer, the guide with vision and passion. As people debate endlessly the pros and cons of standardized content and high-stakes testing, it is encouraging that at least some students get to focus on integrative, real-life projects with significance to the

world around them. In coming years, the students in these stories will be prepared to solve problems by finding data, sifting out the relevant material, integrating varied sources, analyzing relationships, collaborating with others, and communicating their findings. These fortunate students are developing the traits on which the world of tomorrow will depend: scholarship, artisanship, and citizenship.

Charlie Fitzpatrick
ESRI Schools and Libraries

It was a casual remark at the 1999 ESRI International User Conference that sparked *GIS in Schools*. After sitting through the education presentations, the eventual authors remarked to Charlie Fitzpatrick, "There's a book here," to which he predictably replied, "Go for it!" To the entire Schools and Libraries team, Charlie, Angie Lee, and George Dailey: you are the inspiration for this book.

To our teacher writers: you are the pioneers, the visionaries, those who recognize the power of technology to help young people learn in new ways.

Thank you for allowing us to tell your wonderful stories.

To Ali Merayyan and Bill Sutherland: your technical assistance was invaluable. Where our skills ended, yours generally began. We are in your debt.

To Sarah Bednarz and the Geography Education National Implementation Project (GENIP): Your financial support enabled us to meet, hatch plans, and realize our vision for this book.

To Christian Harder, Gary Amdahl, Jennifer Galloway, and the ESRI production team: we gave you rough clay. You shaped it into something worth seeing. Thanks for being so easy to work with.

To those other adventurous educators who might be moved by the *GIS in Schools* stories: your students will be the ultimate beneficiaries of your decision to put GIS technology in their hands. We advise you to also heed Charlie's admonition: Go for it!

Dr. Richard Audet
Dr. Gail Ludwig

Chapter 1

Welcome to Mt. Hope High

Karen Connors' Environmental Science students have been using GIS to supplement their investigations of topics like the water cycle, land use, coastal erosion, and hazardous materials contamination. Connors uses an increasingly popular instructional strategy called Problem Based Learning as a hook to engage students in projects that are relevant and motivating. Her students work in collaborative teams on specially designed problem scenarios that require research, use of sophisticated problem-solving strategies, and applications of technological tools.

A local GIS consulting firm has a school-and-business partnership with her class at Mt. Hope High. With their help, students are learning basic GIS skills and discovering how helpful GIS can be in the exploration of spatial relationships. One team of students has taken a particular interest in a GIS application called OILMAP.™ This GIS-based modeling software integrates weather, tide, and local current conditions to generate oil-spill trajectory predictions that can be displayed over standard nautical charts. These visualizations can guide an entire spill cleanup operation, from the mobilization of emergency response teams to the identification of environmentally sensitive areas that harbor important biological resources.

Capitalizing on her students' interest, Connors developed a study unit on coastal pollution. The GIS consulting firm provided local GIS data and OILMAP software and taught students how to model a spill event. Students completed hands-on lab activities that simulate a small-scale oil spill. From their experiments with the mock spill, they learned the basic principles of containment and dispersion, as well as how nearly impossible it would be, in a real spill, to remove the thick "mousse" of salt water, oil, and air coating miles of shoreline and everything on it, from rocks to clams, seaweed to seals. They studied the chemical and physical properties of petroleum products and the environmental consequences of a major spill. They knew that oiled birds die as often from the toxic effects of ingesting oil as they do from exposure to cold when feathers lose their insulation properties. They found out as well why filter feeders like clams, oysters, and mussels are among the marine organisms most hard hit by oil pollution.

But, until the North Cape Oil Spill happened on a stormy January evening in 1996, no one ever imagined how important knowing GIS would become. On that night a single-hulled barge called the North Cape, being towed by a seagoing tugboat, ran aground and burst a seam in its hull. Thousands of gallons of crude oil were spewed on the pristine barrier beaches of coastal Rhode Island.

The use of GIS in schools depends on the relevance and timeliness of real-world events that act as catalysts.

Images in the local media the next day were powerful: gloved hands shrouding the badly oiled body of a terror-stricken shore bird; a spindrift of dead marine life trapped in an iridescent foamy oil mat; and offshore, a surreal seascape of black breaking waves. Not surprisingly, Mt. Hope High was alive with action on the morning after this dead-of-winter calamity. Students already had a scientific basis for understanding oil pollution and considerable experience using sophisticated problem-solving strategies. Because they were familiar with GIS, students were able to study the spill from a uniquely spatial perspective. And as their inquiries led them deeper and deeper into the problem, traditional lines—between subjects and between teachers and students—were blurred.

Students were in charge. Try to picture the following classroom exchange:

Karen Connors . . . Did any of you ever think when we ran the OILMAP simulation that a major spill would happen in our own backyard?

Dave . . . According to our latest trajectory model, Block Island's smack in the middle of the spill's dispersion path. The slick could arrive there by tomorrow morning!

Matt . . . But the wind has to be right. If it turns south like they're saying, the model's predicting that most of the oil's going to wash up on Moonstone Beach.

Lisa . . . The piping plovers will just love those oil globs!

Dave . . . Forget the plovers, did you see the massive number of dead lobsters that have already started to wash ashore?

Lisa . . . I would never have imagined that this was such a major breeding ground for them.

Matt . . . Hey, can the conversation! We just got a call from the mayor's office in Newport. They're frantic to know what OILMAP is predicting for Newport Harbor.

Lisa . . . Can they boom the entrance to the bay?

Matt . . . No, it's too wavy and too wide. Besides, according to information OILMAP is giving us, most of the oil booms are tied up in that Pennsylvania spill. Maybe they'll use a dispersant or try to burn the oil.

Karen Connors . . . Maybe, but only as a last resort. Let's hope the wind turns before the tide changes. What a disaster!

Lisa . . . You've got to wonder if there's any sense in transporting oil in barges during the storm season.

Dave . . . You won't be the last person to ask that question.

When students sense something real is at stake, learning becomes a natural response.

The actual details of the North Cape spill are real. Shipowners eventually paid a damage settlement of $9.5 million to compensate for the loss of over nine million lobsters, twenty million clams and oysters, five million fish, and countless billions of smaller invertebrates. The high school setting of this scenario, however, is fictitious. It was written to illustrate how students and teachers in special classrooms across the country can and do use GIS technology to investigate spatial problems that are genuine and relevant. GIS is a powerful analytic tool that helps people understand the significance of spatial distribution patterns, whether the issues involve the siting of a new professional sports stadium, animal migratory patterns, or designs for cost-effective school bus routes.

National standards in many curriculum areas support the idea of learning experiences that involve students in what has been termed "authentic inquiry." One of the most direct approaches for stimulating interest is to transform a student's own backyard into a de facto learning environment. GIS is the technology that makes this localization of learning possible.

The purpose of *GIS in Schools* is to illustrate how GIS is being used in actual classrooms by real teachers and students. The case studies in this book are exemplary stories that we hope will encourage teachers who are interested in integrating GIS into their curricula. We also hope these classroom accounts will further motivate teachers who have already begun to experiment with GIS technology.

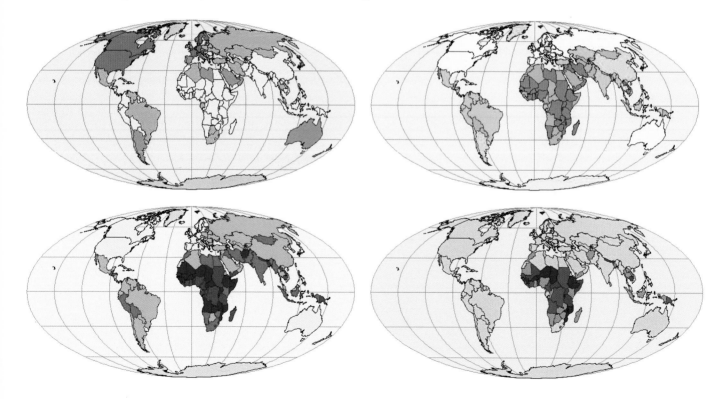

Chapter 2

What is a GIS?

Imagine having your students stack a large number of different maps of the same area neatly in a pile, one on top of the other, and then having them ask questions and find answers using all the information contained on those maps. That, precisely, is what a GIS does. Students can ask questions, sort, and even create new maps from information typically found in atlases, textbooks, the Internet—even on the walls of the school library.

Science generally, and geography specifically, are two disciplines that have benefited greatly from recent technological advances. Information from paper maps can now be transferred to a format that can be stored, analyzed, and reorganized by a computer. Students can get their hands on and use many different kinds and levels of geographic, geologic, and demographic information to help them answer complex questions about the earth and its resources.

- What are snags, where are they found, and how many are homes for nesting birds?

- Where are the "hot spots" for coliform bacteria on the creek and what might be causing this pulse of pollution?

- Where are the cities with a population over 500,000 and what are the geographic factors that they have in common?

- What will need to happen to protect the city if a truck carrying hazardous chemicals crashes on Eastern Avenue?

Questions like these can be answered (relatively) easily and quickly.

GIS combines three words: Geography + Information + System. Geography relates to all the features and processes that occur on the surface of the earth. Information is the heart of GIS, where vast amounts of data are stored and analyzed. Finally, the System is what connects everything—the computer, the data, and a human operator—all working together to ask questions, discover answers, and display them in ways that promote understanding of what it means to live on the earth.

GIS is a way of drawing together vast amounts and different kinds of data and real-world observations into a series of clear and vivid pictures.

GIS is a toolbox for exploring the world

A geographic information system contains specific "tools" that allow a user to question, analyze, and sort through great quantities of data. The heart of the system is a software program linked to a database containing information (for example, population, land-use types, precipitation, or vegetation patterns) that is tied to specific locations. In schools, the GIS software program can allow students to search, move, and query this enormous spatial database and create maps or tables displaying the results of searches. Not only can a person easily find Timbuktu now, they can also make a map showing population distribution in and around Timbuktu, major roads running through Timbuktu, and even agricultural land-use patterns within 100 kilometers of Timbuktu. It sounds easy, and actually is, when the four essential parts of a GIS are properly connected.

An interested student with proper equipment has virtually unlimited access to the world as a complex spatial entity.

Four parts of a GIS

- Hardware

- Software

- Geographic data

- Intelligent user

Hardware

Access to a fast computer with lots of RAM and storage space is the key to successfully using a GIS. The general rule of thumb is: the faster and more powerful your computer, the easier GIS will be to integrate into a classroom setting.

For best results, your computer system should meet these minimum standards:

- Processing speed of at least 233 MHz

- 24 MB of available RAM

- Hard drive with at least 100 MB of free space for data and project storage

- 17-inch or larger monitor with 256 colors

Less powerful systems can still do the job, but the speed of processing map questions (queries) will slow down, and the entire process will naturally take a while longer to complete. Use the fastest and most powerful computer you can find, but don't let these concerns prevent you from starting. The skills and the lessons learned will be the same no matter what kind of equipment you have.

Software

GIS software allows you to sift through all types of geographic information to answer spatial questions. Just as a word-processing program provides the tools for creating and manipulating text, GIS software, such as ArcView® GIS from Environmental Systems Research Institute, Inc. (ESRI), provides many elaborately functional tools for analyzing maps and manipulating data.

These tools include:

- First and foremost, a mapping module for displaying geographic data

- A database module for managing tables of geographic information and a chart module for processing numbers (for instance, population data)

- A page-layout module for creating completed maps and turning tables into charts

- A programming language allowing you to customize the way you question and sort through your spatial information

ArcView GIS combines all these functions into one comprehensive exploration engine.

Like most computer applications, it can take time to learn how to use all the functions available in GIS software. ArcView GIS, however, includes step-by-step lessons using data, maps, and exercises on an enclosed CD that allow even a new computer user to succeed with GIS. Basic features can be easily mastered and will provide a solid base for more complex explorations.

Geographic data

What type of data is needed? In order to get a paper map into the computer, information on the map must be changed into a format that the computer can understand. Since computers only store numbers, the information on the map must be converted into a numerical system by either digitizing or scanning. Digitizing is a manual process that uses a specialized graphics-type tablet and mouse. Scanning is a more automated process where a map is placed on a flatbed or drum scanner that automatically converts the map data to a computer-readable format.

Although this process may sound difficult and time consuming, most GIS software programs now come with huge, ready-to-use spatial databases. It's also possible to purchase ready-made map files for many different geographic regions throughout the world. If you are interested in local data, a city, county, or state often has digital maps they will give to schools. Most states now have spatial data clearinghouses where GIS data at the local level can be downloaded from the Internet at no charge.

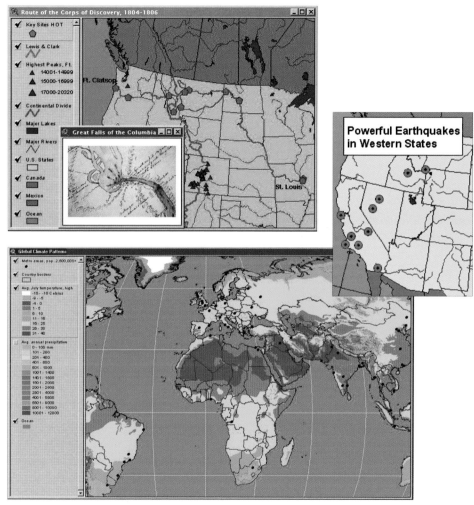

The hard, time-consuming work of digitizing and scanning has largely already been done. Ready-to-use database sets to suit almost any purpose are available and relatively inexpensive—sometimes free for the asking.

Intelligent user

A key to successful use of GIS in the classroom is a user who likes to explore—someone who understands spatial patterns, spatial relationships, and spatial distributions and how they can be used to answer geographic questions. GIS forces you to think spatially—to understand how information can be combined to create a new map or graphic that answers a question, solves a problem, or identifies a new solution to an old problem. It takes someone who is willing to probe, process, and question to make GIS successful in a classroom setting. Students need to be comfortable with exploring, thinking critically, and working cooperatively with others. Teachers need to be comfortable having students following unpredictable and divergent pathways. Using GIS in a classroom can expose students to these essential skills and promote an integrated, collaborative, relevant, and exciting environment for learning. A GIS is not only a means to practical geographic ends, but a learning tool as well.

Your students must be able to spend time working through a progression of lessons with the software and learning how it operates. Mastering GIS is analogous to learning how to drive a car. Successful driving requires learning how to start the car, shift gears, accelerate, steer, and apply the brakes under constantly changing conditions. Similarly, the successful use of GIS in the classroom requires learning how to run the software—shifting from one tool to another, loading databases and moving between them with little effort. Making maps, building queries, sorting data, and printing results are skills that take practice, just like driving. By following the step-by-step lessons on the CD–ROM, students can soon put GIS into action.

Terms and definitions to help you on your way

GIS terms

Spatial data—Spatial data is information about the location of, shape of, and relationships between objects found on the surface of the earth. Spatial data is stored in a database that organizes the geographic information by shape (points, lines, and polygons) and attaches to each item a geographic location (using latitude/longitude or some other coordinate system).

Database—A database is a collection of related information that is stored as a unit. A GIS database includes all the spatial and tabular data for various geographic locations.

Attribute—Attributes are characteristics of a map feature linked to the geographic coordinates of those features. For example, what are the characteristics of a country? What is its population? GNP? Capital? All of these elements, and untold thousands of others, are examples of "attribute data."

Resolution—Resolution refers to the smallest feature in a data set that can be "seen" on a map. If a map has a resolution of 10 meters, only objects 10 meters by 10 meters or larger can be "seen." Anything smaller will not be mapped; anything larger will be mapped. Thus, a 15-by-15-meter building might be displayed on a map, but the 8-by-8-meter building next door would not because it is less than the 10-meter map resolution. The term also refers to the area represented by each pixel in an image, expressed as pixels per inch.

Map—A map is a scale model of some or all of the earth, created by one of many ways of projecting the shape of a sphere on a flat surface.

Map scale—Map scale is the ratio between features on the surface of the earth and features on a map. Real-world objects are reduced by a constant amount when placed on a map, and the map scale tells us the amount of this reduction. A map scale is a ratio, where one unit on the map represents many times that value in the real world. For example, a map scale of 1: 250,000 means that one unit on the map (inches, centimeters, feet, etc.) represents 250,000 of the same units in the real world. A map scale of 1:100 means that 1 inch on the map represents 100 inches on the surface of the earth.

ArcView GIS terms

The list below identifies several important terms needed to understand the operations of ArcView GIS.

Shapefile—Most geographic features can be represented on a map using three basic shapes: points, lines, and polygons. In ArcView GIS, geographic features are generally stored in shapefiles (ArcView GIS can use and create other files). A shapefile consists of a .shp file, a .shx file, and a .dbf file (which should not be separated in user directories).

Theme—A set of geographic features of the same type along with their attributes (location information) is known as a theme. One way to understand themes is to visualize the many different layers of information contained on a map. Each layer of a similar type of data (such as elevations, rivers, cities) in ArcView GIS is called a theme. Themes can be laid on top of other themes to create a single map showing elevations, river locations, and city locations. In ArcView GIS, themes can be turned on and off by a mouse click.

View—A view is basically another term for a map in ArcView GIS. Each view (or map) is made up of layers (or themes) of geographic features, such as rivers, countries, or roads for a particular area or place. Each view can include many different layers of features called themes.

Query—A query is a question asked in a format that ArcView GIS can understand. For example, "How many cities in the world have a population of over one million?" can be easily asked by typing [pop > 1,000,000] in the ArcView Query Builder. If the database being queried contains world city location and population information, ArcView GIS will answer the question with both a table and a map. Queries are among the most powerful GIS tools. They allow geographic questions to be answered that reference all of the data and information contained in a project's database.

Table—Tables are documents containing information stored in rows and columns of text and numbers. Each row (record) represents an additional item (for example, another city or state); each column (field) represents another characteristic about those items (for example, population or climate type). Geographic tables include a set of locational information, along with other attributes. ArcView GIS can read tables in many different formats such as dBASE,® Excel, or text.

Chart—A chart is a graphic representation of the data contained in a table. ArcView GIS can show geographic information in chart format by taking table information and creating a picture of the numerical information. Charts present tabular data in an easily perceivable format.

Project—All the views, tables, charts, and data for a specific topic or area in ArcView GIS are contained in a project. A project is the "file cabinet" for all the materials needed for one activity. Project file names have an .apr extension. A file called river.apr might contain all the information for a huge river assessment. Likewise, a file called wolf.apr would contain the maps, charts, and graphics for a wolf migration study.

Chapter 3

HAZMAT

Walter Paul, Teacher
Chelsea High School
Chelsea, Massachusetts

William Hamilton, Ph.D.
Professor of Geography
Salem State College
Salem, Massachusetts

Over the course of its history, Chelsea, Massachusetts, has experienced its share of troubles, from Revolutionary War battles—a British gun and supply ship was sunk in Chelsea Creek—and devastating citywide fires to municipal bankruptcy. Today, thanks to GIS students at Chelsea High School, residents can sleep a little sounder. This new sense of security stems from a community mapping project developed in my advanced GIS course.

In the fall of 1996, the hazardous materials (HAZMAT) fire chief visited our GIS spatial learning lab while inspecting the new school building, and was intrigued by what he saw happening there. He asked if students could use new Environmental Protection Agency (EPA) software that tracked hazardous chemicals and their locations in the city. He had no idea how vigorous and enthusiastic our response would be.

The number of companies in Chelsea with hazardous materials data registered under the federal Community Right to Know Act (which gives broad powers to communities to act on their own behalf and take control of their resources) has risen sharply in a very short time. In two years it jumped from two to sixty. My students used survey information to generate computer maps of Chelsea's hazardous materials storage sites. Armed with this data and the EPA software, they were able to project the impact of weather, traffic, and other variables on potential chemical spill events.

The GIS class worked closely with the EPA, the Local Emergency Planning Committee, and the Chelsea Fire Department. The culminating event of the project, a simulated toxic spill, involved agencies ranging from ambulance services to the United States Coast Guard. In their central control unit, GIS students analyzed incoming data, made critical traffic-routing decisions, and tracked the spill alongside their emergency management counterparts.

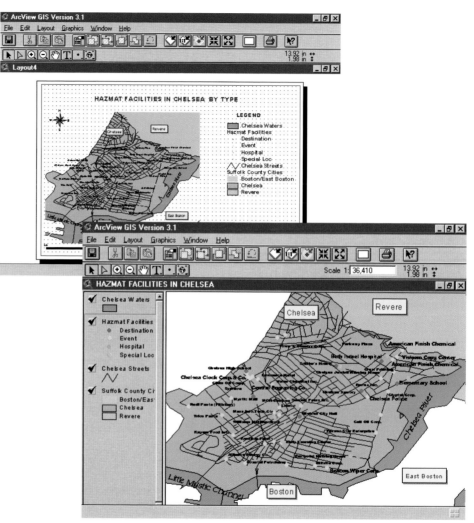

Noting the sharp rise in the number of sites in their community storing hazardous materials, GIS students began the task of, in effect, mapping the risk these sites posed to friends, family, and neighbors.

The project's origin

Several key ingredients contributed to the HAZMAT project's success. The first was that I took steps to develop my own proficiency with GIS technology. I completed two university-level GIS courses and worked for much of the summer developing instructional materials. I incorporated the following goals from an existing geography curriculum that was perfectly aligned with my purposes for using GIS. In this project, students would:

- Work as a team

- Practice problem solving

- Use communication skills

- Know "all aspects" of an industry

- Learn to use math on the job

- Understand Chelsea's resources and community needs

- Produce high-quality work

The use of GIS in schools actually gets students out of school, interacting with mentors, experts, and other concerned citizens and professionals.

The next crucial development was a focusing of the curriculum. With GIS, customization of learning materials by individual teachers is the norm. In my class, students learned basic GIS by using *Getting to Know ArcView GIS,* an ESRI publication that uses case studies to familiarize students with GIS functions. Students made maps of their neighborhoods, city, county, state, and country using ArcView GIS databases. The course emphasized geography as the content and spatial exploration as the skill. There was, however, no unifying central theme for our work. The moment the fire chief walked into our lab, that problem was solved. The course took off, and so did student learning.

The skills necessary to work GIS not only promote learning—they are prized in the job market as well.

HAZMAT at Chelsea High

This course begins with the basics: an introduction to our spatial learning lab, computers and their development, computer operating systems, and a survey of the history of communication technology. Although a vast majority of Chelsea High School's students do not have home computers, their learning curve was rapid. Like most kids today, who are bombarded with technology and sophisticated advertising through television, these students had little trouble using computers to support their learning. The minute they set foot in the GIS lab, they were eager, willing, and able to learn.

Since a primary objective was to have students understand the relationship between data, maps, and how map images are produced, early instruction dealt with cartographic fundamentals such as map scale, projection, distance, and direction. Next, students working in their GIS-supported environment completed activities built on their sense of personal geography. They mapped the journeys their families made to Chelsea from their countries of origin, their morning route to school, and the location of the city's major landmarks.

Students learned about data layers and how digital maps illustrate this information. They created new themes from existing ArcView GIS data. Key map representations (point, line, and polygon) were explored in depth before students advanced to the central focus of the class:

helping the community prepare for emergencies. Throughout the project, teachers emphasized the value of community service as a hallmark of good citizenship.

At midyear, EPA representatives conducted a series of class visits that introduced students to hazardous materials management. EPA officials explained how chemicals and their distribution

have an impact on everyone's lives. Using the Bhopal, India, incident as background, they emphasized the importance of preparedness and explained how an informed citizenry can prevent the loss of life during major spills. They demonstrated how GIS can be employed to prepare for and respond to such emergencies.

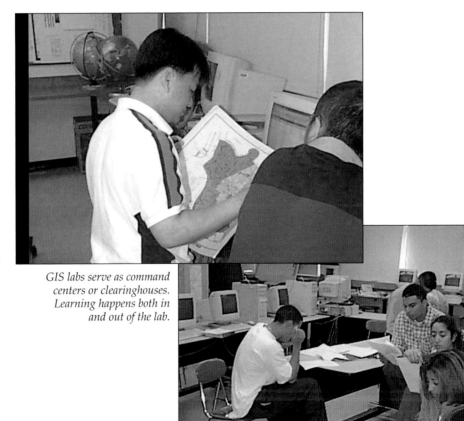

GIS labs serve as command centers or clearinghouses. Learning happens both in and out of the lab.

Trips to the regional EPA office in Boston and to the local power utility illustrated the ongoing role played by GIS in the community. Students reviewed the Environmental Protection Community Right to Know Act before meeting with the Local Emergency Planning Committee. They learned about the city of Chelsea's emergency management plans, facilities, and equipment—and discovered that people with GIS skills are highly employable.

By the midpoint of the school year, students had become well versed in using ArcView GIS and were ready to use EPA's trilogy of HAZMAT software: Computer Aided Management of Emergency Operations (CAMEO); Mapping Application for Response, Planning and Local Operational Tasks (MARPLOT); and Areal Locations of Hazardous Atmospheres (ALOHA). Students were organized into task groups and began to gather local HAZMAT data. Their database included digital photos of facilities, questionnaires completed under law by chemical facility owners and transporters, and floor plans that they drew themselves using Paintbrush. Each task group was assigned to a government agency and was required to submit a weekly work report. The ongoing support of EPA, local emergency management officials, and the Chelsea HAZMAT fire chief was essential for the success of the project.

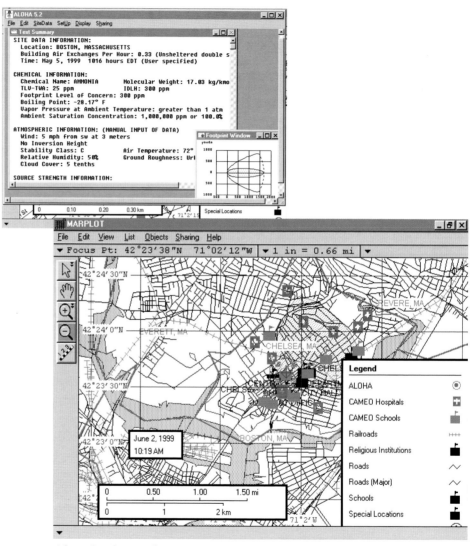

After only a few months spent getting up to speed, Chelsea GIS students were ready to use the EPA's top-of-the-line software.

The HAZMAT simulation

From spring onward, my students worked with the local, state, and federal agencies that would participate in the culminating simulation exercise. The EPA and the students jointly designed a HAZMAT incident that could actually occur in Chelsea—in this case, a collision between a tractor trailer rig and a car, resulting in a spill of anhydrous ammonia.

As part of my curriculum, students were expected to develop an analytic and a spatial perspective of their community and neighboring areas. When a life-threatening event affects a community, dozens of agencies must coordinate their efforts to minimize the consequences of the disaster. Transportation must be rerouted, areas sealed off, buildings evacuated, parents of school age children contacted, hospitals notified, and traffic monitored. With ArcView Network Analyst software, evacuation plans can be quickly determined for immediate implementation.

The cooperation between government agencies, businesses, and citizens helped Chelsea students understand the ways communities work on a practical level.

To complete the simulation of the spill, students needed to develop a set of specific skills. They prepared, studied, and refined the maps and data they would generate during simulation day. They understood the importance of meteorological input, how and where it could be acquired, and how it could affect the disaster. They learned to use wind direction instruments and the sling psychrometer to measure relative humidity. From digital orthophotos, they studied the topography of the Boston metropolitan area and located suitable sites for relocating evacuees. They also identified major routes for moving people to these sites.

On the morning of "simulation day," students were working feverishly at their computers. They arranged for all participating agencies to cooperate, and assumed the roles of officials in agencies they represented. Previously they had blanketed Chelsea with leaflets to alert the community that this was not a real incident.

When the simulation began, students were stunned to find that the EPA had changed the anticipated meteorological inputs. Onlookers were about to get a firsthand look at how well students could use GIS to help them think on their feet. Gas dispersion and puddle models indicated new areas of concentration, and predicted plumes billowing in different directions. Using Network Analyst, Chelsea students rose to the occasion, finding new evacuation routes and safe places for fleeing people to head for. Officials on the scene were highly impressed by the ease and accuracy with which the students conducted the simulation.

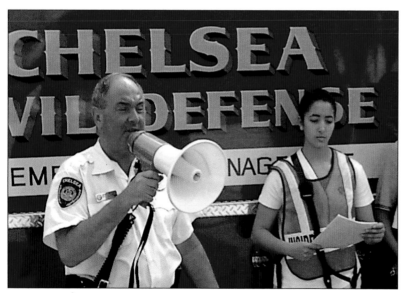

By changing one of the preset conditions of the simulation, EPA officials forced students to think on their feet.

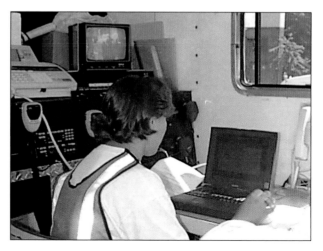

Today, students in seven adjoining communities have integrated this successful educational project into their curriculum. Through HAZMAT, GIS persuasively demonstrates its adaptability and usefulness and shows the community at large how it can help students gain geographic and scientific knowledge and technological skills. Today, everyone sleeps better in Chelsea, particularly my hard-working students who learned a vivid lesson about how young people can contribute to the health and welfare of their community.

Using GIS in schools has a double benefit: students become excited about learning at the same time they begin to understand the interdependent web that a community actually is. The community is strengthened by every inquiring step a student takes.

Chapter 4

Eyes on wildlife

Becky Rennicke, Teacher
Prairie Wind School
Perham, Minnesota

In the early 1990s, wolves returned to Perham, an event not anticipated by Minnesota's Wolf Management Plan. When state and federal agencies put out a call for more information about these canine visitors, Perham teachers saw an opportunity to introduce authentic research into the seventh- through twelfth-grade science curriculum. The Eyes on Wildlife Project, which began in 1995, originally targeted the wolf populations in Minnesota, but has since expanded to include research on deer, bittern, and black bear populations.

The project incorporates manual and technology-based wildlife research methods such as visual tracking and scent post surveys; radio, aerial and satellite telemetry; the Global Positioning System (GPS); and geographic information systems (GIS). Local, state, and federal data sets provide the base against which students build and compare their own data. Students then make this information available over the Internet.

Wildlife research in schools

A major goal of conducting genuine wildlife research is to have students become directly involved in ecosystem management. Students who develop an intimate and personal knowledge of wildlife populations, who share this information with their community, and who see their work having an impact on public policy, gain an appreciation of their role in, and responsibility for, the environment. A classroom project like Eyes on Wildlife makes research and wildlife management clearly relevant to students' lives.

Grant monies and cooperative agreements with research biologists enabled project participants to purchase radio and satellite collar equipment for tracking the animals. Students witnessed the capture of "their" wolf and in some cases assisted in collaring the animal whose behavior they would monitor. These hours of active work in the field, alongside professional researchers, observing animals with whom they could feel a kind of kinship, encouraged strong feelings of personal investment in the project. Students returned to the classroom having made important connections to the practical realities of the field, and were consequently better able to determine what additional work needed to be done. These tasks included library research on wolf ecology, and data mapping and analysis—work for which the students alone were responsible, and which they completed successfully.

Wolves give new meaning to the idea of hands-on experience. This tranquilized animal is being fitted with a sturdy tracking collar.

Regularly revisiting field sites with telemetry equipment, students determined the exact whereabouts of their wolf by homing in on the signal produced by the collar, taking compass readings, and obtaining accurate geographic coordinates using triangulation techniques or a GPS unit. They then logged that data and obtained additional information from field biologists. Complete reports were sent via e-mail to other project participants who used them to analyze their animal's movements.

Subzero temperatures did not keep students from the field. Here they use radio telemetry to track collared wolves.

Wolves watching people watching wolves

Beginning with an original question posed by a biologist or student, team members worked their way toward answers, coming to a dynamic understanding of animal behavior and wildlife ecology while learning about the importance of research and field data, developing visual tracking techniques and skill in using telemetry equipment to gather scientifically accurate information. They conducted library research on the ecological niche of wolves and were able to compare this information to field data about their animal's food and habitat preferences.

The high point came when students began to put the pieces of their puzzle together. Each investigation involved a spatial question associated with the dispersal and migration patterns of an animal. What is the animal's preferred habitat? Does the animal interact with people? Is the animal found where human population densities are high? Do the animals travel along roadways? What is the range of an individual animal? What is the average distance that animals travel in a day? Does the time of year affect migratory patterns? All of these questions were investigated by students participating in the Eyes on Wildlife Project.

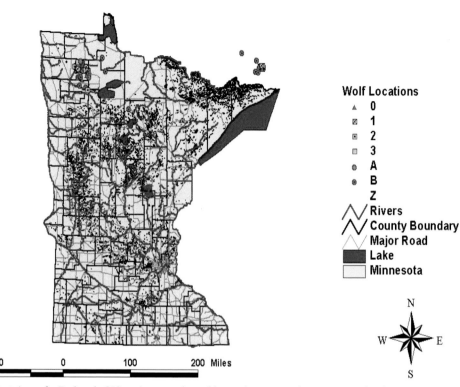

The trigger for Perham's GIS project was the sudden and unexpected appearance of wolves near town. Students began with two simple questions: Where did these wolves come from, and where are they going?

Problems and . . .

When the project began, all data was superimposed on USGS topographic maps. Students plotted animal locations and color-coded other types of data such as date and time of day. These initial maps were messy, and each research team conceded that there seemed to be no practical solution for handling the massive volume of field data spewed out by the transmitting collars. Our students remained undaunted.

Dedication resulted in success, but our successes resulted inadvertently in problems. One issue dealt with predicting subsequent locations of animals that wandered over large distances. When an animal struck out in a new direction we were concerned about not having full map coverage. To do a complete analysis we needed broad access to topographic maps and aerial photographs to analyze their paths. If we purchased these maps ahead of time, what would happen if the animals traveled in an unpredicted direction?

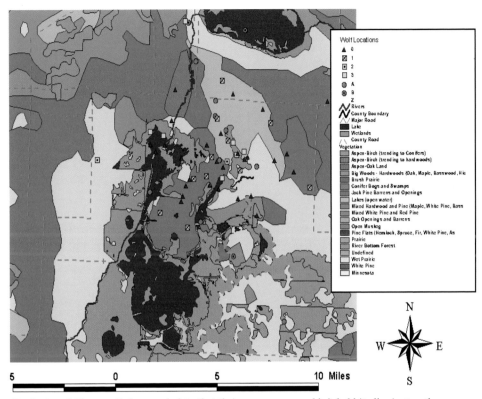

Students quickly compiled so much data that their paper maps couldn't hold it all—just as the areas covered by the maps were too small to hold the wolves.

Another problem arose when both the number and kind of animals being tracked increased. In addition to wolves, field research began to investigate black bear, deer, and bittern populations. More collared animals multiplied the number of issues to be managed.

Field research was conducted at two different sites: the Agassiz National Wildlife Refuge in northwestern Minnesota, and Camp Ripley in central Minnesota. Both areas cover many square miles of territory. Paper maps presented a severe limitation for the type of research the project had undertaken. These expanded operations increased the operational and research concerns. We found ourselves asking how much map coverage we could afford. We were being forced to select between animals to be studied and those to be ignored, and to concentrate only on the information we believed ahead of time would hold answers to our research questions.

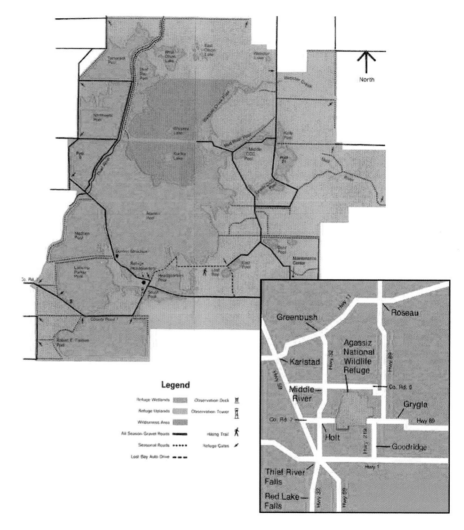

Legend

Refuge Wetlands	Observation Deck
Refuge Uplands	Observation Tower
Wilderness Area	
All Season Gravel Roads	Hiking Trail
Seasonal Roads	Refuge Gates
Lost Bay Auto Drive	

Field research was conducted over vast areas: wolves require a lot of elbow room. Eventually, research expanded to include other species, such as the bittern (left).

. . . Solutions: GIS and Eyes on Wildlife

As luck would have it, several people associated with the project attended an introductory GIS workshop and saw the power of this technology. They realized at once that this technology offered a potential solution to the project's problems. GIS enabled the research team to map all animals. Each of the maps we needed was available in digital form for all of the animals being tracked throughout their migratory ranges.

The project team has never had second thoughts about incorporating GIS technology into Eyes on Wildlife. We have found that students as early as the seventh grade can successfully plot locations, color code time of travel, and overlay topographic and aerial photos. When students manipulate their data with GIS they get an instant visual display of what would otherwise have been less clearly meaningful tables and rows of numbers. Migratory patterns and relationships among variables jump out of the maps. Students begin asking their own questions and seeing how GIS can help answer them.

Kids often learn best when information is presented visually. The pictures and brightly colored, easily generated maps are what students—especially in the early going, as they learn their way around the technology—appreciate most about GIS. At their command, maps appear on the computer screen with new data displays and color patterns for them to analyze and decipher. Students are encouraged rather than forced to make decisions about what data is relevant for answering questions, and what isn't. You might even say they're eager to take on these kinds of responsibilities.

Eyes on Wildlife has only begun to realize the full potential of GIS software. As more is discovered about the applications of this powerful technology, its use as a tool for learning and conducting original research will expand. In our project, GIS is providing important clues that reveal the mysterious secrets of animals in the wild—an interesting and important connection for students to make between nature and science.

Chapter 5

Jump-starting GIS: Youth and teachers as colearners

Bob Loudon, Director
San Bernardino National Forest
San Bernardino, California

Carol Ann Franklin
University of Redlands
Redlands, California

In the San Bernardino Mountains of southern California, towering trees, thick shrubs, boulders, scurrying lizards, prosperous squirrels, and, of course, standing dead trees create a living classroom where young people prepare to become caretakers of public lands by learning about forest ecosystems. The program is Children's Forest (CF), a collaborative project between Children's Forest Association and the U.S. Forest Service, in which students engage in fieldwork that sharpens skills in communication, critical thinking, decision making, and teamwork—at the same time practical real-life research is going on.

"So is that a snag?" asks the tired teenager squinting in the hot sun as she points to a dead tree. Most of the kids in the group are wondering if learning about snags is worth the scratches and sweat they're working up from hiking for an hour through thorny bushes and rugged terrain. "Remember, we're looking for standing dead trees that are at least 2 meters tall and 25 centimeters in diameter," the group leader reminds everyone. "It looks like this one qualifies," says the tired teenager.

The small group—three teenagers and three teachers—begins collecting information on the snag using various measuring and data-gathering tools, which include a Trimble Pathfinder Pro XL GPS unit. They collect data on height, diameter, species, decay class, and location. The group will eventually use the information to determine the quality of this important wildlife habitat. Snags are an often-overlooked component of a dynamic, healthy forest, providing important home sites for insects, mammals, and birds. Raptors (hunting birds) use them for perches, woodpeckers make cavities in them with their sharp beaks, and secondary nesting birds depend upon unoccupied cavities for their nests.

Dead trees are not useless trees, as this group of GIS teachers and students is about to learn.

Today is the first day of an intensive week-long workshop in which classroom teachers have been teamed with young people to learn a technology that will help them study the forest. A natural partnership emerged: the mission of Children's Forest is to involve youth in stewardship projects, while the University of Redlands has a historical commitment to helping teachers integrate new technologies into their teaching. And because GIS is such an effective tool for displaying and analyzing spatial data, this software became the technological focus of the workshop. To involve both students and teachers with GIS, CF and the University of Redlands instituted "Jump-Starting GIS," a collaborative teacher education program in which teachers and children become colearners.

The project relies on close curriculum connections with Children's Forest, incorporates new strategies that link students and teachers as learning partners, and explores opportunities for introducing GIS into the K–12 curriculum. Working in collaborative teams, members learn to use GIS software, collect field data, and determine how to integrate this type of activity into the classroom. Participants include students (ages 13–16) and area teachers (upper elementary, middle, and high school).

Day one

On Monday morning, participants are introduced to their team partners and begin mapping a section of the University of Redlands campus. Both teachers and students are a little apprehensive about working together. Coming into the workshop, teachers assume that they will have to take on traditional classroom leadership roles, but they quickly discover that this will not be the case. By week's end, most will be grateful to be working as equals aside their considerably younger teammates.

Team members soon become experts in cartography, as they learn the importance of measurement, scale, and compass use. Teams regroup to develop a layered map in the old-fashioned way by adding their data to a basemap with pens, pencils, and acetate sheets. Their final challenge of the day is to assemble all sections to create a comprehensive map of the campus.

Their exploration of GIS as a tool for spatial analysis is now about to begin.

The "Jump-Starting GIS" workshop began with mapping expeditions around the University of Redlands, a project cosponsor along with Children's Forest.

Day two

After delving into mapping basics and procedures for environmental analysis, the time is ripe for an introduction to ArcView GIS software. Learning partners complete a combination of ArcView GIS exercises from *Getting to Know ArcView GIS* and activities built upon Children's Forest data. This is the first real test of the student/teacher relationship. Teachers struggle to learn to work GIS software beside students who are sometimes quicker to grasp concepts than their adult partners. Working through the basics of ArcView GIS occupies this full day and half of the next.

Day three

The workshop moves to the mountains. After an orientation to Children's Forest given by students who are trained "Children's Forest Snagology" leaders, the teams use GPS units to capture and record snag data. When they locate a snag, they collect data about its location and information about diameter, decay class, height, and use by wildlife. As part of the workshop experience, each team completes an additional field research project, collecting a variety of data, such as the location of used nesting boxes, trail markers, and vegetation distribution. Throughout the project they monitor a number of different information sources to capture any social, biological, and physical environmental data that might be significant in their analysis and problem solving.

Hard-copy mapping gave way on day two to instruction in mapping software. Data collection began on day three.

Day four

Today the teams huddle around computer screens at the University of Redlands, looking at their "snagology" data. They are visualizing their field data and beginning a preliminary spatial analysis in ArcView GIS. The data they generated in the field is part of an inventory of standing dead trees in the Children's Forest. Their data is added to a data set that includes additional information, such as elevation, roads and trails, vegetation, water sources, and land use. Workshop participants are poring over the forest basemap to determine its habitat suitability for particular wildlife species, analyzing the distribution of snags, and asking questions about the size and distribution of specific tree types.

This is the time when GIS learning connections are made. Participants are using their newly acquired ArcView GIS skills to analyze their own data and the larger data set. Real questions quickly and naturally emerge. Group dynamics are particularly interesting during this stage, as students challenge discouraged teachers to persevere. Because they often have more advanced computer skills, students teach teachers, who happily become students. On the other hand, the teachers foster adult behaviors among students, and encourage them to explore alternative problem-solving approaches. Through this experience both teachers and students gain more confidence in and respect for their respective strengths and abilities.

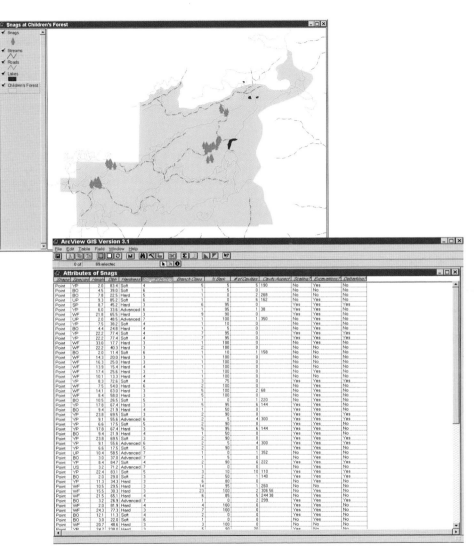

This table suggests the scope of the data collected in the mountains. The screen capture above shows one layer of data graphically represented.

Day five

Teams return to the campus basemaps they created on the first day, and with their newly acquired GIS skills, are challenged to solve a real-world problem. In the first year of the program, the task was to identify the best site for a new science complex; in year two, the task is to locate the least disruptive path for fiber-optic cabling of the campus. During the teams' presentations, campus decision makers are in the audience to compare simulated findings with the actual decisions that have been reached. This feedback proves invaluable to participants in its reinforcement of the importance of working with GIS to solve problems.

Day six

On the final day, participants apply their GIS skills to topics that are limited only by their imagination and interest. The spatial analysis projects use online GIS data sources and information available on the ESRI K–12 CD–ROM. One team looks at the proximity of science museums to urban areas, while another develops a "lost ship" activity, and a third analyzes the frequency of earthquakes with respect to school sites in the county. For their final workshop activity, teachers and the youth participants share ideas for future GIS-enhanced school projects.

Climbing mountains, collecting data, learning new computer and GIS skills, analyzing, presenting findings: the workshop was an endurance run that demonstrated the amount and quality of learning that happens when GIS and the real world meet in a classroom.

Tired but happy

The Snag Workshop offered participants the opportunity to learn about maps, spatial analysis, environmental modeling, data collection and management, GPS, and applications of ArcView GIS for displaying information. By the conclusion of the workshop, participants had completed an environmental modeling and analysis project, collected and analyzed snagology data, developed a spatial analysis project with a colearner using GIS data from several sources, and designed a classroom project that would incorporate GIS.

To help assess the kinds and quality of learning that went on in the workshop, participants completed pre- and post-work questionnaires that examined their understanding of GIS and problem solving. Each person responded to writing prompts that dealt with the effectiveness of each day's activities. Upon completion of the workshop, teachers were also asked to reflect on the readiness of their school districts to implement GIS, on both the practical and academic level. Overall, though overwhelmed at times by the quantity and intensity of the learning that the brief workshop had afforded, all participants seemed eager to take that learning back to their classrooms.

The snagology workshop balanced hard days of fieldwork in rugged conditions with intensive classroom sessions where students presented work to real-world decision makers.

Chapter 6

Nonpoint source pollution: An assessment of impervious land cover

Kathryn Keranen
Thomas Jefferson High School for Science and Technology
Alexandria, Virginia

Imagine almost 400 square miles of heavily populated coastal plain, with rolling hilly uplands and steeply sloping valleys. This is the landscape of Fairfax County in Northern Virginia. In colonial times, the county was mostly prime agricultural land. Today, it is a thriving business and residential area, the most populous and wealthy county in Virginia. Yet Fairfax County, which borders the District of Columbia, still boasts over 30,000 acres of open-space parkland.

According to the most recent census, the county has a population of almost one million, a median age of thirty-six, and an average of 2.7 persons per household. During the next century the population is expected to increase with a corresponding rise in the number of housing units.

A cursory examination of conditions suggesting explosive growth prompted students in our senior-level Geosystems course at Thomas Jefferson High School for Science and Technology to begin their study of land use and nonpoint source (NPS) pollution. TJ, as we like to call it, is a public school serving twelve hundred students who are admitted on the basis of high aptitude and interest in pursuing careers in mathematics, science, and technology. Geosystems is a senior-level science course presented as a program that integrates all prior science experiences under the banner of topics such as geology, meteorology, oceanography, and astronomy. GIS and GPS are the key integrating technologies that students use in their Geosystems projects.

Fairfax County, Virginia, is a complex, fast-growing, and fast-changing part of metro D.C.

Nonpoint source (NPS) pollution

As rainfall or snowmelt flows downhill over the surface of the earth, it captures a variety of natural and man-made pollutants. Eroded sediment from construction sites, oil and grease from road deposits, and fertilizer and insecticides used around the home are common examples of NPS. Some of this contaminated runoff is deposited in wetlands such as lakes, rivers, and streams; other portions can contaminate groundwater. Thus, these pollutants have a detrimental effect on habitat for wildlife, recreational amenities, and drinking water supplies. The EPA has identified NPS as the leading cause of water quality problems.

NPS is almost completely anthropogenic. That is, it results from human activity. Through practices like erosion control, conservative fertilizer and herbicide application, and proper handling and disposal of harmful household chemicals, much of this load of pollutants can be reduced. Students at TJHSST also learned that the amount of impervious surface (parking lots, roads, buildings) is a prime factor in estimating the threat of NPS to water quality.

Anderson Classification Scheme

1	Water
2	Bare earth
3	Cleared ground
4	Crop/Field land
5	Transitional vegitation
6	Native forest
7	Coniferous forest
8	Wooded residential (20% impervious)
9	Low-density residential (20–35% impervious)
10	Middle-density residential (35–50% impervious)
11	High-density residential/trans (50–65% impervious)
12	Low-density commercial (less than 80% impervious)
13	High-density commercial (greater than 80% impervious)
14	Structure (100% impervious)

The type of ground cover dominating in a given area has a significant effect on that area's ability to handle pollutants.

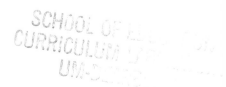

Impervious surfaces

The Geosystems project used GIS to look at the effect of impervious surfaces on the watersheds of Fairfax County. The amount and distribution of impervious surfaces is closely linked to nonpoint sources of pollution and can serve as indirect indices of water quality. An impervious surface is one that is impenetrable by water. Imperviousness measures the percentage of water flow that is able to penetrate the land surface. For example, the surface of asphalt roads and man-made structures is 100-percent impervious. Noncoated surfaces vary in permeability depending on factors such as slope, vegetation cover, and soil composition. For the purposes of this study, measurements of imperviousness were limited to paved surfaces and rooftops. The most accurate method for making estimates of imperviousness is to actually visit a site. Since our study area covers such a large area, direct observation was impractical. Instead, we used satellite imagery to evaluate the landscape.

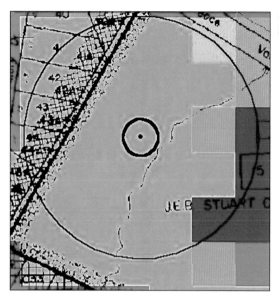

Ground truth point with land-cover theme.

Ground truth point with roads and parcels theme.

Earth-orbiting satellites

The U.S. government began launching earth-monitoring satellites in the 1960s to supply data for resource managers, scientists, and political scientists. Today, NOAA is responsible for operating the Landsat system, maintaining the USGS Earth Resources Observation Systems (EROS) archives, and distributing Landsat information. In a Landsat image, each pixel covers a 30-by-30-meter area. Color values correspond to a variety of USGS land-use categories that can be interpreted using specific classification schemes.

Landsat data was the primary source of information for the impervious surface analysis completed by TJ's Geosystems students. Our first discovery was that computers make mistakes interpreting satellite data. Commercial buildings were sometimes classified as pastures! To check for the degree of discrepancy in the computer's interpretation, we sampled a set of random points scattered across the county and compared actual land use to what the computer was saying. This method, sometimes called ground truthing, allowed us to determine that the computer's classification of regional land cover was 78 percent accurate. This ground truth protocol enabled us to assign an accuracy assessment to the land-cover images and adjust them accordingly.

TJ students used data displayed by ArcView GIS to distinguish among features such as water, cropland, coniferous forest, and low-density residential areas. Since land-use type can be correlated with estimates of imperviousness, and the degree of imperviousness is directly related to water quality, analysis of corrected satellite imagery provided us with the means for assessing the past, present, and future impact of pollution from nonpoint sources.

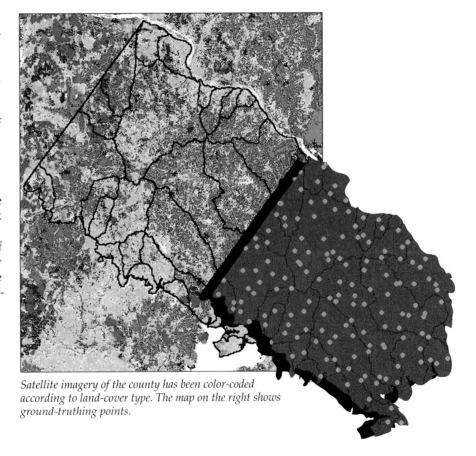

Satellite imagery of the county has been color-coded according to land-cover type. The map on the right shows ground-truthing points.

Growth and Fairfax County's watersheds

Fairfax County's unprecedented growth during the 1980s and 90s is readily observable by conspicuous alterations that have been made to the natural land-scape. Regional satellite images confirmed these changes and provided a basis for predicting future land-use trends.

Overall, our study showed a 15-percent countywide increase in impervious area between 1987 and 1993, the period for which data was available. Changes were particularly noticeable for high-density commercial and high-density residential areas. When broken down by individual watersheds, values ranged between 3 and 50 percent.

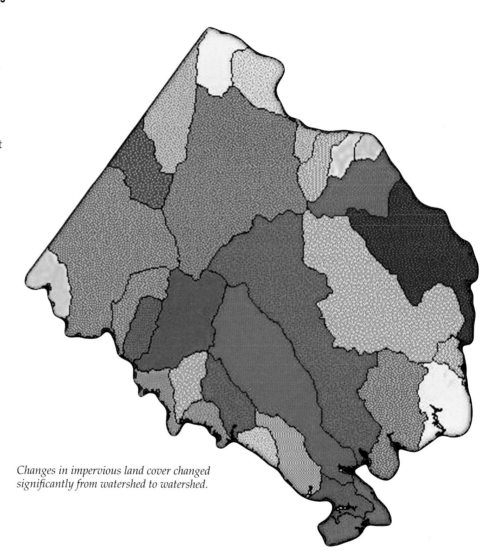

Changes in impervious land cover changed significantly from watershed to watershed.

A watershed includes the area drained by a river and all of its tributaries. Fairfax County has thirty-five watersheds. Groundwater here is shallow and moves eastward from higher elevations toward the Potomac River, which runs through our nation's capital. Since precipitation cannot penetrate an impervious surface, both runoff and the amount of NPS increase when land is developed. Our study examined the relationship between changes in the amount and location of impervious surfaces and stream discharge in these watersheds.

By calculating the percentage of impervious area at various time intervals for each watershed, we created a watershed change index. Fortunately, the USGS maintains several continuous monitoring gauges in Fairfax County that measure daily stream discharge. We used data from these stations to construct hydrographs that show fluctuations in stream flow before, during, and after storms. When there is a significant change in impervious land cover there is a corresponding change in the shape of the hydrograph. Thus, our analysis of stream discharge allowed us to infer the impact of changes in impervious surface on water quality. We found that both impervious area and stream discharge increased over the seven-year period covered by the data.

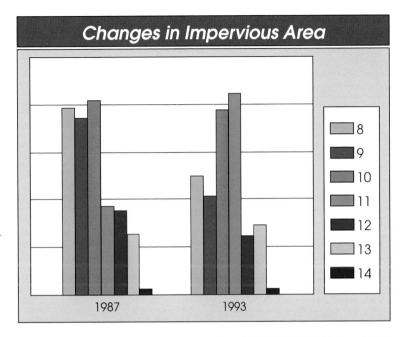

Imperviousness Error Matrix

GIS Mapped	Not Impervious	Low Intensity	High Intensity	Total	Error
Not Impervious	44	5	5	54	0.185
Low Intensity	5	24	1	30	0.200
High Intensity		3	1	4	0.750
Total	49	32	7	88	
Error	0.102	.25	0.857		

Implications

Many things happen to a raindrop during its journey to the sea. Largely uncontaminated water droplets or snowflakes move through the many familiar pathways of the water cycle. The students in the Geosystems class at TJHSST analyzed how development in their county, as measured by the increase in impervious surface, was affecting stream runoff. They found a direct correlation between impervious area and stream discharge. GIS was an important visualization tool that helped them to reach these conclusions. The GIS maps that students produced were the focal point of their public presentations of their findings.

Because it is part of Metropolitan Washington, D.C., development in Fairfax County appears to be a foregone conclusion. Through their integration of Landsat imagery, stream-flow data, and GIS, students produced a study and a set of powerful recommendations for reducing the impact of future development on nonpoint sources of pollution.

Chapter 7

Historical documentation of a culture

Ann Thompson, Teacher
Ligon Middle School
Raleigh, North Carolina

Marsha Alibrandi, Assistant Professor
North Carolina State University
Raleigh, North Carolina

Rita Hagevik, Teacher
Ligon Middle School
Raleigh, North Carolina

As buses were pulling away on that last day of school in 1998, a huge bulldozer knocked down two 60-foot willow oak trees to make way for a new building. "If only trees could talk!" lamented one of the teachers. Those trees had stood as living monuments to J. W. Ligon High School, which opened in 1953 and was the premiere black high school in Raleigh, North Carolina. Ligon attracted the city's best and brightest black teachers, many of whom were educated in the nearby historically black colleges. Located in a traditionally African American neighborhood, Ligon became for a time the center of the African American community and a great source of pride. In 1971, at the height of desegregation, Ligon became a junior high school in the consolidated Wake County Public School System. Today the building is the home of a magnet middle school.

For years, an active alumni association from the African American community has kept the memory of the high school alive. A feeling of resentment, symbolic of all that was lost during desegregation, continues to be associated with the building that now houses the magnet school.

The felling of the huge willow oaks, symbols of stability and permanence, mobilized teachers and spurred collaboration with Ligon alumni to document the school's history. Teachers were optimistic that lessons from these rich historical accounts would guide the future of the new school and restore its status as a community center and source of pride for alumni. They also hoped that students would be able to draw strength from the school's rich cultural tradition and appreciate more profoundly their heritage and place in the Ligon community.

Above: the felling of the huge, old willow oaks that symbolized the strength and integrity of the Ligon school's heritage was a blow to the whole community. Here students and teachers nurture a new beginning. Right: J. W. Ligon (1869–1925), pastor, principal, publisher, was fired from his position at Crosby–Garfield School for "daring to run for public office."

Ligon History Project

Ligon High School alumni, the school's teachers and students, university faculty, graduate students, and community partners collaborated on what became known as the Ligon History Project. The project's mission was clear: identify, locate, document, and showcase the remarkable chronicle of Ligon High School. Middle school students in journalism, history, science, and GIS classes would recreate a "virtual" history of the school integrating traditional research methodologies such as interviewing, data collection, and archival research with state-of-the-art technologies like the Internet and GIS. The overwhelming scope of this task quickly became apparent, when it was discovered that many records and archives from the original school, like those of other large urban districts, had been either lost or destroyed during the desegregation period.

Ligon students participating in the history project became familiar with the technology of geography and history on several levels: from paper maps to GIS software and satellite imagery.

The goal of the Ligon History Project was to immerse students in a real-world learning experience that included historical data, maps, and archival records. Middle school teachers Ann Thompson and Rita Hagevik, along with North Carolina State University professor Marsha Alibrandi, agreed that GIS would be an ideal tool to support the project. GIS maps of Raleigh during the 1950s, 60s, and 70s were prepared that told the story from an African American perspective. The project coordinators' initial challenge was to determine:

- How GIS could be used to represent historical events

- How an undocumented history could be graphically represented using GIS for later display on the Internet

- How GIS could integrate quantitative data with archival and qualitative interview data to tell a story

Aerial photography provided another perspective of the Ligon community for students to consider.

The teachers involved in the history project, from left: Ann Thompson, Ginny Owens, Rita Hagevik, Neville Sinclair, Bettie Mackie.

Two grants supported the highly motivating field trips that kicked off the project. At the City of Raleigh's GIS office, students learned about the powerful adaptability of GIS. They observed state legislators using this technology to weigh different scenarios before reaching complex spatial decisions. When students were shown an annexation map illustrating the growth of the city, they realized that GIS was a tool that could help them analyze and illustrate changes that occurred in the Ligon area from the late 1700s to the present.

At the State Archives, students studied historic maps and photos that depicted similar evidence of the city's growth. Students were particularly impressed with an 1800s bird's-eye photograph of Raleigh taken from a hot air balloon. Sanborn Fire Insurance maps from 1949 showed clearly defined "colored" or "Negro" areas of Raleigh. The maps were shocking to many students, but vividly evocative for alumni who had lived under segregated conditions. Witnessing these maps helped students to understand that the history of the segregation period from an African American perspective has not been recorded as "official" history. This single event more that any other gave students the motivation to preserve and communicate the history of this period.

Later, on the Great Raleigh Trolley Tour, students visited places they had studied using GIS and maps. What had formerly been simple spatial representations took on life and meaning when students explored the historic areas of their city.

Raleigh Annexations

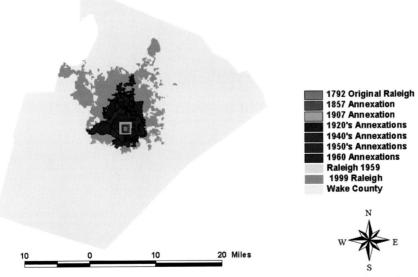

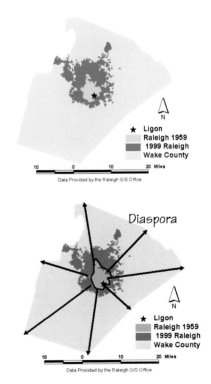

One of the goals of the Ligon History Project was to understand the community's place within greater Raleigh as the city expanded and changed.

GIS's role in the Ligon History Project

Enthusiasm is the word that best describes the project's kickoff. Students had armfuls of maps provided by the State Archives and a set of base layers for their GIS database. Annexation boundaries, Wake County streets, and downtown and southeast Raleigh land parcels were made available to the project by the Raleigh GIS office. Wake County GIS provided census tracts and public school data. A sense of panic, however, developed when it occurred to students and facilitators that all of this data represented Raleigh in just one year, 1999. Information overload was as much of a stumbling block as were information gaps. A new action plan emerged that scaled back the original scope of the project.

The project involved gathering historical data and learning new GIS skills. Students read transcripts of oral history interviews of alumni conducted by journalism classes. They analyzed old school yearbooks and local history books. They generated a list of buildings that were important to Ligon's history. Graduate students from the local university gleaned data from the State Archives. GIS students learned interview techniques that they used when speaking to elder alumni. The intergenerational aspect of the project had a magnetic effect; students were learning history from those who had lived it.

Though a good deal of time was spent on computers acquiring GIS skills, the history project sent students just as often out into the world, collecting data and observing life from many different viewpoints.

Life map

We coined the term "life map" for the eventual product: a historical representation of Raleigh in the 1950s. The map included:

- Comparative spatial interactions between cultural groups

- Layers of data for gaining alternative spatial perspectives

- Archival documentation that validated the African American experience

- Identification of landmarks for historic preservation

Students completed two additional GIS projects using historical maps provided by the State Archives. They created basemaps for a Railroad Development GIS project and a GIS Greater Raleigh Trolley Tour and Web site project. Photos of historic buildings were linked to these GIS maps.

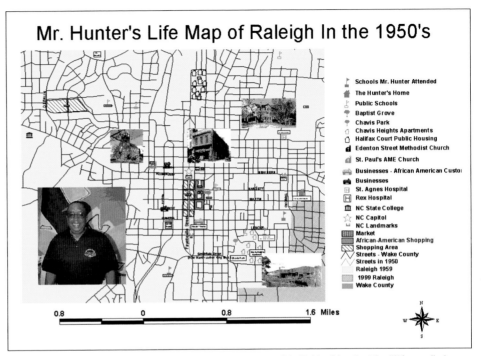

The Ligon History Project works on both the community and individual levels. The "life map" above graphically and spatially illustrates one man's journey through life and neighborhood.

Sharing history

At the end of the year, students organized a community exposition that enabled them to reconnect with the alumni who inspired the project. Students, parents, teachers, university professors, and graduate students also participated. Even the governor attended. The students erected stations where visitors could view exhibits of the three GIS-based projects: the Ligon High School History Web site; yearbooks, the book of oral histories, and *Capturing the Past to Guide the Future* (a videotape of selected interviews); and architectural history reports. Professors, teachers, and students described the collaboration and shared future plans for the project. These include an on-site school history museum and an Earth Quilt that will become a stop on Raleigh's Millennium Trail. Visitors were impressed by the sophistication and quality of the students' interdisciplinary displays.

The exposition was the capstone of the project. Teachers and students worked diligently to prepare exhibits that effectively communicated their project. The spirit and synergy of this intergenerational, multicultural community service group drawn together by their common interest in Ligon inspired everyone connected with the project. GIS provided a dynamic medium for communication. For the first time, people saw how history could be successfully mapped, documented, and vividly illustrated using GIS.

The year-end exhibition of the results of the students' work was one of the most important elements in the history project: the story of a community had to be shared with the community.

Chapter 8

Investigating an urban watershed: How healthy is Deer Creek?

Bob Coulter,
with Nate Litz and Nathan Strauss

Litzsinger Road Ecology Center of the
Missouri Botanical Garden
St. Louis, Missouri

Two former students of mine, Nate Litz and Nathan Strauss, conducted an investigation of the Deer Creek watershed, an area of approximately 40 square miles that drains many of the small cities near St. Louis, Missouri. As their fifth-grade science fair project, the students planned to measure the water's health as it meandered downstream, assess changes at different points in the watershed, and explain those changes. Clearly, GIS software would play an invaluable role as a tool for displaying and analyzing information about Deer Creek.

After conducting a preliminary survey based on safety and access issues, I presented Nate and Nathan with several potential testing stations. They chose sites evenly spaced along the creek that were also likely to contain flowing water throughout the project's four-month duration. Since Deer Creek tends to dry up seasonally in its upstream reaches, they started at the midpoint and tracked water quality from there. They chose three locations along Deer Creek and one along the River des Peres. The last site was selected to allow for a comparison of water quality between the end of Deer Creek and the river into which it drains.

Nate collects water samples while Nathan pinpoints the test site location with a GPS receiver.

Water quality index

To track changes over time, Nate and Nathan scheduled four test dates between November and March for data collection. Under my supervision and guidance, the boys used a variety of testing equipment, including Vernier probes and Texas Instruments' Calculator Based Laboratory (CBL) equipment, chemical test kits from LaMotte, and EasyGel coliform tests from Micrology Laboratories. By measuring nine environmental parameters, they were able to calculate a water quality index (WQI) for each site, on each test date.

While there was some variation in individual measurements, the overall water quality for the sites showed a considerable decline in water quality at site 3. However, the water quality "recovered" after the creek drained into the River des Peres—a possible indication that the pollution was being diluted in a larger volume of water.

This dramatic change in water quality raised several intriguing questions. Was the pollution from a point source, such as a sewage leak? Could a nonpoint source such as bank erosion or storm runoff be responsible for the deterioration in water quality? Or, was the sudden drop in water quality due to a combination of both of these factors? Background research using resources in the Missouri Botanical Garden library and an interview with Tony Lafferty of ESRI's St. Louis office began to shed light on their findings. After completing a brief tutorial on GIS and learning how to add themes, zoom in and out, and label features, Nate and Nathan were ready to use the software to help them explain their data.

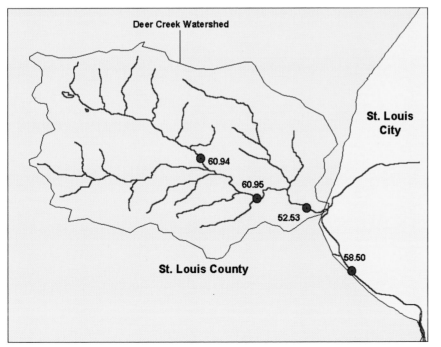

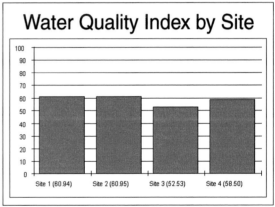

Average results for four months of testing show a significant decline at site 3.

Investigating sources of pollution

In their search for potential point sources, the students analyzed maps containing zoning data provided by the St. Louis County Planning Office and environmental data provided by the Missouri Department of Natural Resources. The latter data set identified the organizations having permits to discharge effluent into Deer Creek and its tributaries. Four of the eight permitted sites shown with red dots on the accompanying map were found between site 2, where water quality was highest, and site 3, where it was lowest. Two of the permitted sites were located directly on Deer Creek, and two discharged into Black Creek, a tributary that flows into Deer Creek downstream from site 2. Generally, the zoning patterns show a marked change from residential (green) to commercial and industrial (red and purple) downstream near site 3.

After reviewing their water quality data and land-use maps, the boys concluded that a major point-source problem might be industrial pollution. For example, the limestone quarries upstream from site 3 could account for the rise in pH and Total Dissolved Solids, since it is quite possible that carbonate ions were part of this discharge. These two factors (pH and Total Dissolved Solids) combined to account for 23 percent of the drop in the WQI.

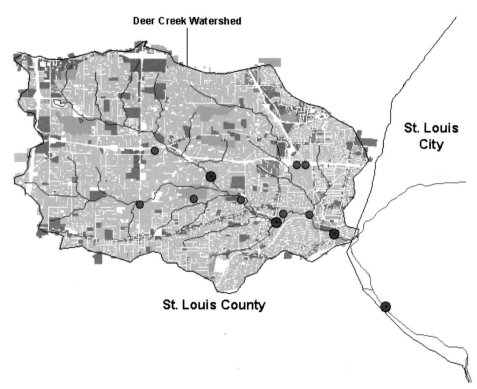

Land-use patterns in the watershed suggest one possible explanation for the decline in water quality.

The boys also decided that a second point source could be a sewer leak or discharge. Like many cities, St. Louis uses a combined sewer system for storm runoff and untreated sewage, which can overflow into creeks during periods of heavy rainfall. The maps suggested that this might be occurring between sites 2 and 3. A brief lesson on bacterial contamination helped the boys understand that the rise in fecal coliform levels that they recorded could explain the higher biochemical oxygen and the lower levels of dissolved oxygen in the water at site 3. Nate and Nathan calculated that these three factors

(fecal coliform, biochemical oxygen demand, and dissolved oxygen) accounted for 63 percent of the drop in the WQI.

They knew from their background research that nonpoint source pollution, such as that caused by storm runoff and bank erosion, could also contribute to lower water quality. Site 3 is immediately downstream from a large shopping center parking lot with large amounts of runoff that includes sand, dirt, and oil. Also, stream channeling to accommodate a shopping center left exposed banks 20 feet high that likely contributed to

the higher turbidity levels recorded at this site. This rise in turbidity accounted for the remaining 14 percent decline in the WQI.

As a result of this study, the boys concluded that a combination of industrial uses of the land, the design of the region's sewer system, storm runoff, and bank erosion all contributed to Deer Creek's low water quality. Without a single culprit to blame, stream recovery will be difficult. The chart below summarizes the results of Nate and Nathan's investigation.

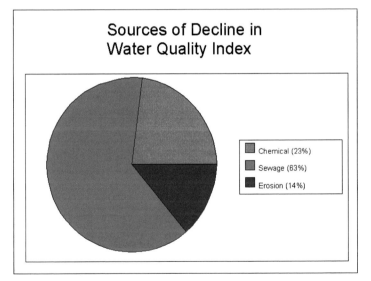

Sources of Decline in Water Quality Index

- Chemical (23%)
- Sewage (63%)
- Erosion (14%)

Nate and Nathan concluded that several sources contributed to the lowered water quality at one of the test sites.

Further investigations

This project raised many interesting issues about the health of the creek. Nate and Nathan measured a significant change in the water quality downstream and linked these changes to land-use patterns. Future student work could build on this project by testing for specific forms of pollution. For example, a calcium carbonate analysis would help to determine if limestone is affecting water quality. Further testing might be able to pinpoint where fecal coliform levels rise, indicating a possible sewer leak. The influence of storm runoff and bank erosion on water quality at site 3 could be measured by testing immediately before and after storm events.

Other questions raised by Nate and Nathan's data had to be left unanswered because of equipment limitations. One particularly interesting study might focus on the rise in total dissolved solids (TDS) between November and March. Since this rise coincides with the winter storm season, it is possible that this change could be caused by road salt running into the creek. This winter a number of student groups will be using chloride testing equipment to study the relationship between road salt application and water quality.

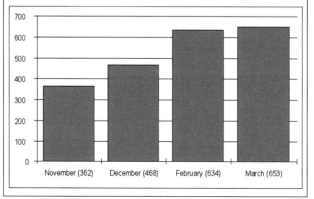

The increase in dissolved solids during winter months suggested a need for further investigations of the watershed.

Increase in Total Dissolved Solids

| November (362) | December (468) | February (634) | March (653) |

Why GIS?

There are many water-monitoring projects being conducted by school children and environmental organizations. What makes our project noteworthy is its use of geographic information system (GIS) software by elementary age students. The maps created by Nate and Nathan allowed them to see the spatial relationships that were critical for understanding the environmental condition of Deer Creek. Because of its unique ability to integrate data in tabular form with graphs and maps, GIS enabled Nate and Nathan to identify the locations where they collected data and to analyze the relationships between water quality and land use. This type of analysis can open new possibilities for students in their development as critical thinkers. Today, as part of the Missouri Botanical Garden's Environmental Monitoring Program, more than one thousand students are replicating Nate and Nathan's pioneering work. The health of the Deer Creek watershed may ultimately be determined through the research efforts of these remarkable fifth graders.

Nate, Nathan, and their teacher Bob Coulter worked together to make a difference in the welfare of their community.

GIS in nontraditional settings

Reinhold Friebertshauser, Teacher
University School
Hunting Valley, Ohio

Nine years ago, Terry Harmon, my colleague at University School, and I first learned about GIS. We immediately recognized its potential as a tool to integrate his outdoor science teaching with my computer graphics lab. Our challenge was to develop appropriate ways to introduce GIS into our curricula. We have since discovered that this problem is a common one when teachers are intent on bringing GIS technology into K–12 school settings.

We made our first attempt in an earth science course that had a strong field component. Three teachers and fifteen students met for two hours every day. Students measured the campus with tape rulers, mastered the survey transit, tested soil samples with homemade soil kits, digitized maps, cataloged vegetation layers, created digital elevation models (DEMs) and digital terrain models (DTMs) for our pond's watershed, and hunted for local GIS data.

We quickly learned some painful lessons. Because almost no K–12 schools and few professionals in our area were using GIS technology, we were essentially on our own. Learning was through trial and error . . . many trials and many errors. We found that some of the techniques we thought we had invented were virtually identical to those being used in professional settings. Our first lesson, then, was to trust our instincts and ignore our failures.

We discovered that collaborative projects were the most effective approach for using GIS. Originally, we tried to assimilate GIS into the traditional framework of a textbook-bound school world. The first task assigned to students with GIS could just as easily have been completed without the technology. We now conclude that we should have been using GIS as a tool for solving difficult but genuine problems. Our early actions were in contradiction to our vision of a GIS-enhanced educational experience. In effect, we were "collecting garbage with a Cadillac" because we were unable to liberate ourselves from a tradition-bound approach to teaching and learning.

Former University School teacher Pat Barnes stripping native brook trout eggs with Vince Laconte of the Ohio Department of Natural Resources. The eggs are hatched as part of the University School's hatchery program to restore the Brookie to Ohio streams.

Location of the Chagrin River Watershed.

Chagrin Watershed Institute

In 1995, no one was actively studying problems in the Chagrin River. This watershed covers a 250-square-mile area east of Cleveland that eventually drains into Lake Erie. The Chagrin River cuts through a landscape of wide fields and deep valleys that expose a dramatic glacial history. Small towns like Chagrin Falls and Gates Mills, dating back to the Western Reserve settlement of the early 1800s, dot the valley. Century-old homes rest among modern subdivisions. Vegetation in the watershed is represented by classic northern hardwood forest. Many sites along the Chagrin have Scenic River designations from the Ohio Department of Natural Resources. Because it accepts drainage from forty different municipalities, this beautiful landscape, with its gentle Appalachian foothills, cold, clear streams, and diverse wildlife, is threatened by urban sprawl.

To expand our GIS applications, we founded the summer Chagrin Watershed Institute, a soil and water conservation project. Financial assistance comes from a private foundation. A competitive process provides fellowships for projects that address community problems associated with the Chagrin Watershed. The charter members of the program consisted of four teachers from neighboring public schools and fifteen of their students. This team completed four weeks of coursework to develop the GIS skills needed to study the watershed.

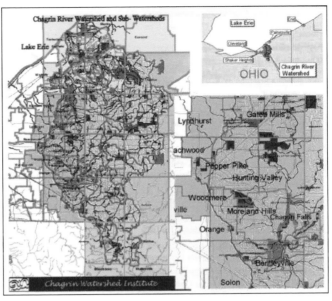

Composite of watershed location, ArcInfo™ coverage of watershed, and a zoomed-in view of ArcInfo coverage showing a qualitative habitat evaluation index (QHEI) data table for colored 600-foot beaches of stream, as well as the area for a proposed soil erosion control project.

Aerial photograph of the upper school campus of the University School.

Since its origin, the Institute has provided free GIS services to organizations. We regard ourselves as "educational and service-oriented opportunists" who search for purposeful work that serves the community. Our GIS skill base enables teachers, students, and community members to collaborate as partners. Over its five-year history, the Chagrin River Institute has been involved in a variety of watershed initiatives.

- A feasibility study for using Landsat satellite imagery to quantify impervious surfaces in the Chagrin Watershed. With support from the U.S. Environmental Protection Agency, the analysis was completed for the Chagrin Watershed Partners, a nonprofit consortium of cities and towns dedicated to solving watershed-based problems.

- A zoning map project by Advanced CAD/GIS students at University School for the Geauga County Auditor's office. These digital maps are now part of the county's public information access system.

- A small subwatershed delineation study of the Chagrin River basin by student summer fellows that refined work done by the Ohio EPA. Chagrin Watershed Partners and other area organizations make extensive use of this database.

- An environmental monitoring effort that used digital data-gathering equipment to study environmental variables such as stream flow, turbidity, conductivity, and soil moisture. GIS-based hydrologic models integrate this data.

- A man-made wetland that provides tertiary treatment for the school's sewage treatment effluent. Students use this site as a learning station.

- A two-year program for building a seven-county digital open-space database for the Cleveland Metropark system. This database represents the only regional open-space GIS coverage for the area.

- The only public-access GPS base station in northern Ohio. With equipment donated by Trimble Navigation, our station can correct GPS data to determine latitude/longitude positions with 1- to 3-meter accuracy. More than forty organizations use this service, including the Ohio EPA and the National Park Service.

- A pilot test of an integration between ArcView GIS and the Ohio EPA's Qualitative Habitat Environmental Index (QHEI) to evaluate stream quality. The team designed low-cost bench-top Calculator Based Laboratory (CBL) systems that can be used in the field to monitor environmental parameters.

- An extensive digital library for the Chagrin Watershed that displays soil type, slope, zoning districts, demographics, impervious surfaces, streets, open space, land parcels, vegetation, and developed areas. A digital database was built that offers photographic documentation of sites within the Chagrin Watershed.

- A successful challenge to a sedimentation control plan by a local municipality associated with channeling headwaters of a pristine trout stream. Students and teachers argued before the city council with support from the U.S. Army Corps of Engineers and Cleveland Metroparks. Construction was delayed until more environmentally friendly and cost-effective alternatives were investigated.

- A trout behavior and habitat restoration project that relocates brood stock from the only known population of spawning native brook trout in northern Ohio. This work includes a partnership consisting of the Ohio Department of Natural Resources, Cleveland Metroparks, the Geauga County Park District, the Cleveland Trout Club, and Trout Unlimited.

These uncommon types of educational events happened when interested teachers and students from different schools collaborated on local and meaningful projects.

Issues affecting the GIS in education world

GIS work in the Chagrin Watershed Institute depends upon technical skills that student fellows learn in my Computer Assisted Design classes over a one- to two-year period. Teachers in our project would certainly affirm the importance of what philosopher Mortimer Adler calls didactic learning—lectures in which the teacher conveys information—but the Institute creates, as well, places and times in which the other two elements of Adler's learning triad thrive: coaching, in which students are challenged by teachers in more individualized, interactive settings; and Socratic learning, in which teachers and students engage in intellectual dialogues.

In this kind of environment, the combined skills of the group provide the richest resource. Everyone is both a specialist and a generalist. Team members who were viewed as GIS or GPS specialists waded into the streams they mapped. Those who were oriented toward political activism struggled to master GIS to strengthen their arguments. Other members became adept at analyzing stream water quality. Some became land-use experts.

University School students and Chagrin Watershed Institute participants installing a pressure transducer data logger to measure stream flow in one of the main branches of the Chagrin River.

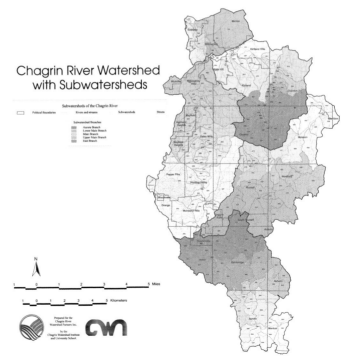

Chagrin River Watershed with Subwatersheds

Because the Chagrin Watershed Institute is nontraditional, we wonder if more traditional school settings can support similar experiences. When a public hearing deadline has to be met, our team members often worked ten to twelve hours straight. Can traditional classroom settings replicate problem solving like this? School day schedules, even those with two-hour blocks, hardly lend themselves to such project work. Terms such as "authentic," "hands-on," and "problem-based learning" are meaningful only when linked to actions that produce real outcomes for students.

Cutting-edge technology is seductive. If we assume that good teaching and learning necessarily flow from applications of good technology, we are ignoring two thirds of the equation. As a powerful analytic tool that enables students to solve real problems, GIS offers this same attraction. But is a student with a GPS receiver making appropriate use of this technology by spending thirty minutes watching digital readouts of longitude and latitude coordinates? Our project, because of its collaborative, skill-based, task-oriented approach, contains checks and balances that mitigate against such seduction. We believe in the importance of giving students access to advanced technology only if they are taught to use it wisely.

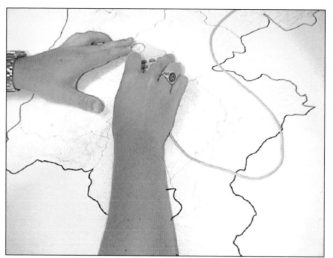

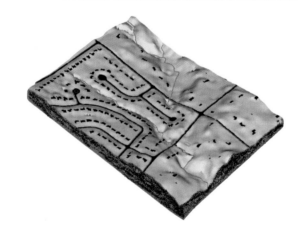

A student digitizing hand-delineated topo sheets into GIS (left). Once the data is digital, it can be used to create models, such as the digital terrain model of the proposed soil erosion control project (below).

The potential for GIS in schools is profound. In our program, teaching GIS technology was not the goal, but rather a means to an end. It was one of many types of glue that bonded our cooperative endeavor. For teachers who choose to integrate GIS into their curricula, tenacity and resourcefulness, along with software support from companies like ESRI, are necessary for getting the job done. To use GIS, students and teachers have to develop a level of competence that rivals the skill level of the organizations with whom they collaborate. They also have to be conversant in other fields that are connected to their GIS work, such as chemistry, math, geology, ecology, and economics.

Our GIS journey at University School began with the question, "How do we get GIS into our curriculum?" The attempts we made to answer it led us down a less than fruitful path. But when we asked, "How can we, as K–12 teachers and students, use GIS technology to do things that are truly useful to our community?", we transformed that dead end into a world of broad horizons. Our project no longer concerns itself with formal connections between GIS and the curriculum. A far more profound movement is afoot in the Chagrin Watershed Institute. Individual teachers and students sneak GIS technology into the curriculum because they need this tool in the context of collaborative endeavors. And they do so in ways that work for them.

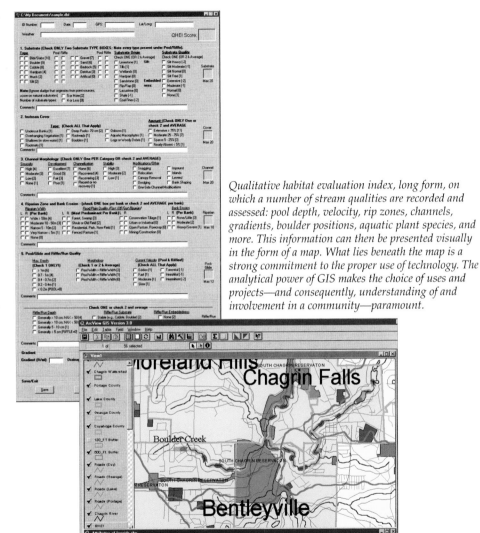

Qualitative habitat evaluation index, long form, on which a number of stream qualities are recorded and assessed: pool depth, velocity, rip zones, channels, gradients, boulder positions, aquatic plant species, and more. This information can then be presented visually in the form of a map. What lies beneath the map is a strong commitment to the proper use of technology. The analytical power of GIS makes the choice of uses and projects—and consequently, understanding of and involvement in a community—paramount.

Acknowledgments

This is a revision of an article that
appeared in the April 1997 issue of
GeoInfo Systems. The author thanks
the organizations and individuals that
generously supported our efforts,
especially Jack Dangermond and
Charlie Fitzpatrick of ESRI, and the
University School for giving teachers
the academic freedom to engage in
such pursuits.

For more detailed information about
these projects, visit the Chagrin Watershed Institute's Web site at cwi.us.edu.

Chapter 10

Getting GIS into the hands of kids: Overcoming obstacles

Kate Dailey, Principal
Bishop Dunne Catholic School
Oak Cliff, Texas

Bishop Dunne is an urban, coeducational, middle and high school in Oak Cliff, a community fifteen minutes southwest of downtown Dallas. The school enjoys an ethnic diversity that mirrors the neighborhoods that surround it. Established in 1961, Bishop Dunne has weathered years of problems: financial straits, deteriorating facilities, a poor academic reputation, and threats of closing. By 1995, Bishop Dunne had reached its nadir. Enrollment had fallen to critical levels, educational technology meant typewriters and overhead projectors, and school debt exceeded a million dollars. In 1996, I became its principal.

My priorities were to raise academic standards, update our technological resources and integrate them in a meaningful way across the curriculum, increase professional development opportunities for teachers, and generally reenergize the school. I wanted Bishop Dunne to be a great school, and I wanted everybody in it to want it to be a great school, too. I had to build a team of staff and faculty who shared a vision, a supportive school board and Diocesan administration, and people from the community willing to help.

The school's reputation improved quickly and dramatically. In 2000, Bishop Dunne was named as one of the top Catholic schools in the nation, in recognition of the ways we have brought technology into our classrooms and effectively used it. The school's enrollment has risen to more than five hundred students and is still growing. More than 35 percent of the students of each graduating class are the first in their family to receive a high school diploma, and 98 percent of our graduates continue on to college.

The interest in people and places that's inherent in the study of geography is also a key element in revitalizing the natural desire of students to learn.

Technology and its connection to school improvement

The Bishop Dunne GIS story is concerned less with GIS itself and more with technology's role in supporting academic excellence. From the beginning, we believed that technology was the key for improving conditions at Bishop Dunne.

Before coming to Bishop Dunne, I taught at the K–8 St. Monica Catholic School, also in Dallas, where I discovered GIS through my association with the National Geographic Association and ESRI. Intrigued by the technology and the idea of using GIS as a way of getting more students more fully engaged in their learning, I infused GIS technology into geography and science courses. The importance of using relevant and real community issues as the focus for student investigations and the application of technological skills is quite clear. Students should leave our schools with both the desire and the ability to investigate the personal and community questions that will surely arise during their lifetimes. Helping them gain technological proficiency as well as decision-making and critical thinking skills can only enhance their development as informed and involved citizens.

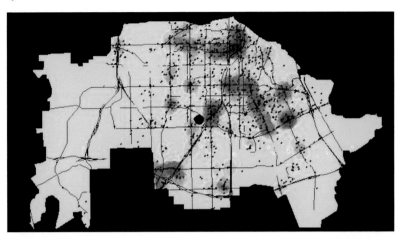

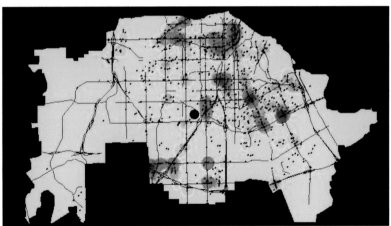

One of many projects undertaken by students in Bishop Dunne's geotech research lab was a series of crime analysis maps that compared robberies and aggravated assaults in 1993 and 1999. Using ArcView Spatial Analyst and the Crime Analyst extension, they discovered significant reductions in nearly all the hot spots.

The first project that St. Monica's students tackled was a local reservoir that was dying. In the White Rock Lake undertaking, students conducted science experiments and community interviews, built GIS maps, and made a presentation to the Dallas City Council. The project was spurred by a community need to preserve an important resource. An ancillary benefit was that students experienced how powerful scientific and technological tools contribute to research and decision making.

These two photographs suggest the scope of student inquiry, from aerial photogrammetry to hands-on fieldwork.

Technology and its connection to school improvement

At Bishop Dunne, the amount of capital we could raise through tuition failed to cover school expenses; finding the means to acquire the hardware and software we needed became a major problem. Our Technology Committee's dream was to have five computers in every classroom and three state-of-the-art computer labs.

Finding and securing community support was the first initiative. Because we had a coherent and practical vision, and had doubled enrollment in just four years, companies and foundations were quite willing to step in. Our school's record in receiving grants is quite remarkable and today amounts to well over a million dollars in outside funding. Seventy percent of this money has been spent on technology.

We built our first lab with matching grants from two Dallas foundations, and realized almost immediately that one lab wouldn't be able to meet surging student demand. Within three years, additional grants were obtained that enabled us to equip two more labs and introduce five computers into 50 percent of all classrooms. The school was fully rewired, cabled, and equipped with a T1 line.

From digitizers (left) to digital cameras (below), Bishop Dunne's commitment to excellence has always underscored the need for students to become familiar with the new tools of computer technology.

Improving the academic profile of any school is a continuous process of examining and reexamining curriculum and instructional approaches. Professional development is a key ingredient of reform. At our school, all teachers are regarded as technology teachers. Following a master plan, specific technological experiences for students are woven into each discipline's course framework. Systemic integration of technology calls for careful collaboration between administration and faculty. Strong incentives have been instituted for teachers to encourage their own technological education. By contract, teachers must belong to professional organizations and engage in thirty hours of annual professional development. An annual stipend of one thousand dollars and the title Technology Master Teacher are awarded when a person completes a rigorous program of skill building.

Skill building is not for students only. Incentives and contractual agreements for work above and beyond the call of the classroom encourage and support professional development.

Geography, education, and academic excellence

Visitors to our school see maps. They adorn every school wall. Geography is a required course for all students at Bishop Dunne. It fosters a divergent way of thinking about and seeing the world, and in doing so, supports the school's mission. When visitors look beyond our maps, they find that spatial literacy is embedded throughout the school's curriculum. Geography links other subject areas, too, particularly biology and environmental science.

The adoption of geography as a unifying school theme was an important element in the blueprint for reform at Bishop Dunne. The plan was closely allied with GIS technology, the integration of which began as we developed awareness among our teachers. The potential for using this interdisciplinary tool to foster spatial and analytical skills among students was readily apparent. Specific GIS courses began for interested science and geography teachers. As more teachers became intrigued and excited with GIS, additional professional development opportunities were provided. In 1998, the school hired a GIS teacher, created a GIS lab, and developed three GIS-supported courses that were linked to social studies and science.

Teachers today have a heightened awareness of what technology in the classroom can mean to student learning. Lectures are uncommon. Active student involvement in learning dominates. Group projects, electronic portfolios,

student-directed inquiry, and community partnerships are all in the mix and contributing to the new sense of what education means at Bishop Dunne Catholic School. GIS is the centerpiece of that experience.

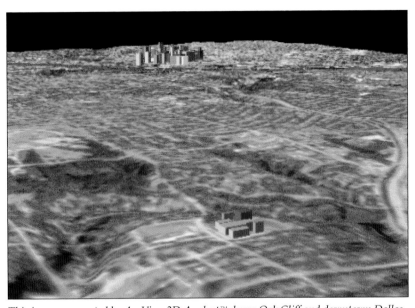

This image, generated by ArcView 3D Analyst,™ shows Oak Cliff and downtown Dallas, looking from the southwest to the northeast. One of the most challenging of the Geotech lab projects, it is also one of the group's crowning achievements.

The GIS curriculum and its teacher

Brad Baker, a teacher renowned in the Texas Geographic Alliance for his commitment to geographic education and the use of technology in the classroom, heads the GIS program. The original GIS curriculum was written by me with the help of ESRI personnel. Brad expanded the program and made it his own. He applied a two-pronged approach by forming specific GIS classes, and by creating conditions that supported the infusion of GIS across the entire academic program. Many courses now deal with their subjects from an integrated and interdisciplinary perspective. Our GIS course offerings have grown into a three-course sequence.

Brad is a natural teacher, one whose enthusiasm and passion for his own learning is passed along to his students. His philosophy is that education must extend beyond the school building so that students can become active members of the community. Brad's students use ArcView GIS to support their research-driven and community-based projects. Dunne's Environmental Science class, in partnership with the EPA and Texas Parks and Wildlife Commission, completed a study of a local creek and its water quality, ecology, and biology. Another group of students analyzed car theft data for the Dallas Police Department and identified high-risk areas.

Working with the Dallas Fire Department, students mapped historical airplane crashes around Love Field. With a local commercial realty company, they are creating an interactive Las Colinas Business District mapping program.

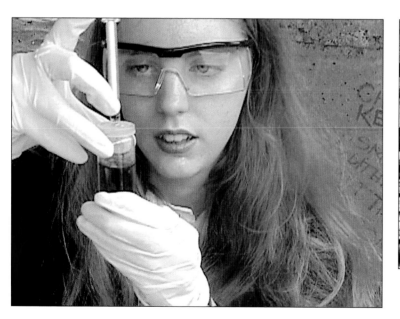

Testing water quality on the Trinity River and on Five Mile Creek. Concerns about the pollution and its effects on public health drove a number of student projects.

Conclusion

As we assess the turnaround of Bishop Dunne Catholic School, we find that the most important elements were unflagging determination, faith and trust in our mission, a firm belief in educational reform and integrated real-world teaching and learning, a dedicated faculty willing to explore areas beyond their comfort zone, a supportive administration, and students who are excited about learning. It is paramount that our students, who come from a range of socioeconomic backgrounds, have every opportunity to cross the digital divide and become proficient with the emerging technologies of the 21st century. That is precisely what makes the story of Bishop Dunne so compelling. It proves that an economically distressed school can form a vision and an ambitious plan to bring technology and GIS into every classroom, and thereby transform itself into a place where learning really happens.

Kate Dailey, Principal
kdailey@bdhs.org

Evelyn Hinkle, Assistant Principal
ehinkle@bdhs.org

Paul Wood, Technology Director
prwood@bdhs.org

Brad Baker, GIS Teacher
babaker@bdhs.org

Chapter 11

Toronto: Geomatics and three hundred thousand students

Imagine six or ten stadiums filled with students. Imagine a small city of students, population more than a quarter million. Now imagine all those students querying data tables and creating new GIS layers in a study of the way their school district is run, or their city, their country, their world.

That's what happened in Toronto.

The 1999–2000 school year marked the beginning of a wave of school reform throughout the province of Ontario. Though some requirements of "Bill 74" have proved controversial, changes in the curriculum have generally been welcomed. One of those changes emphasizes the importance of understanding the expanding role that technology plays in education, business, industry, and government. All subject areas must now include some kind of practical technological experience within the frame of coursework. The mandate for the ninth-grade-level geography course, for example, calls for the integration of course materials and one or more geotechnologies, a term the Ministry of Education uses to describe all new technology relating to geography, including remote sensing, GIS, the Global Positioning System, surveying, aerial photography, and cartography.

The momentum created by the positive response to this reform and widespread support from teachers prompted officials in the ministry to make an extraordinary decision: to license ArcView GIS for distribution to all publicly funded schools in the province.

With help from ESRI–Canada in the form of software, data, and technical support, the ministry has been able to provide access to GIS (or geomatics, as it's known in Canada) technology for more than three hundred thousand students, nearly seventeen thousand teachers, and one thousand administrators. Apart from the benefits intrinsic to GIS-enhanced social studies, the initiative bridges as well the gap that too often lies between tools and techniques used in classrooms and those of the marketplace.

"The fact," said Allan Hux, District Wide Coordinator for Social, Canadian, and World Studies, "that the Ontario government . . . is giving students access to ArcView GIS shows a strong commitment toward the future of social education at all levels for years to come." What might have become just another technology dump—replacing organic learning with the building of simple technical skills—has become instead a way of understanding the relationships between global forces and daily lives, between the economies of nations and the change in a student's pocket, between the fish in the sea and the fish on the table.

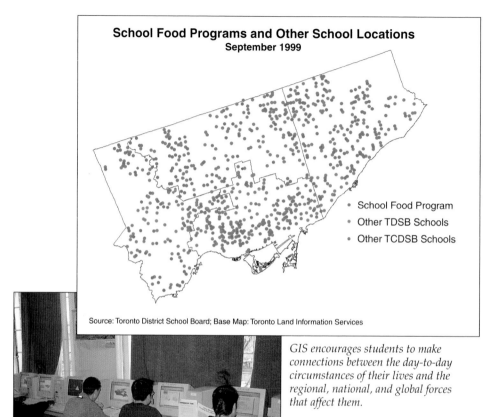

School Food Programs and Other School Locations
September 1999

- School Food Program
- Other TDSB Schools
- Other TCDSB Schools

Source: Toronto District School Board; Base Map: Toronto Land Information Services

GIS encourages students to make connections between the day-to-day circumstances of their lives and the regional, national, and global forces that affect them.

The School District Site License

The City of Toronto was amalgamated in 1998 from six municipalities. The newly configured city covers 632 square kilometers and has a population of two-and-a-half million, making it the fifth largest city in North America, smaller than Chicago but larger than Houston. The simultaneous merging of seven boards of education resulted in the creation of the Toronto District School Board (TDSB), the largest board in Canada. The TDSB manages a budget of over two billion dollars. Six of the ten Canadian provinces, by comparison, have smaller budgets than the City of Toronto alone.

TDSB entered into a three-year agreement for a GIS School District Site License (DSL) aimed specifically at the K–12 community. This arrangement allows for unlimited classroom and administrative use of ArcView GIS, ArcView Spatial Analyst, ArcView Network Analyst, ArcView 3D Analyst, and ArcIMS.™ The license also provides unlimited access to all ESRI K–12 data sets and automatic software upgrades. ESRI–Canada makes available as well a Canadian data CD and offers training and technical support to all the schools in the district.

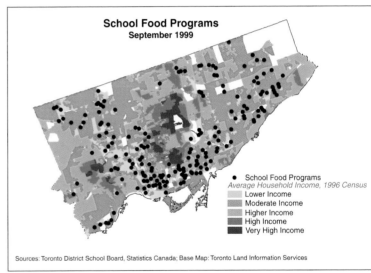

Given the power to examine the world around them from spatial and demographic perspectives, some students in Toronto focused on their own school district's policies.

Train the trainer

A key element in the province's plan to use its GIS as fully and effectively as possible is a professional development program for educators. TDSB and ESRI–Canada jointly created a series of training opportunities built on the "train the trainer" model. In the first year of the program, 120 Toronto secondary school educators learned the basics of ArcView GIS. These people in turn shared their knowledge with colleagues at their own schools. During the second year, elementary school teachers were included in the basic course, while advanced users became acquainted with the software's extensions. The third year will focus on making multidisciplinary connections. The ultimate goal of the program is to achieve districtwide competency in geotechnologies.

The goal in Toronto is to have everybody—from students in grade schools to district administrators—become familiar with geotechnologies.

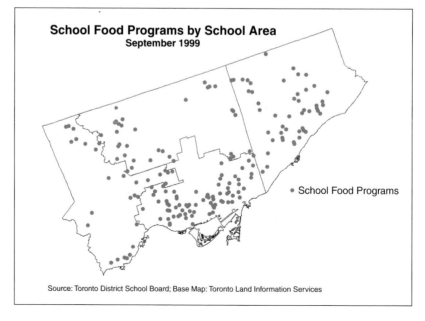

School Food Programs by School Area
September 1999

• School Food Programs

Source: Toronto District School Board; Base Map: Toronto Land Information Services

Toronto District School Board as an enterprise GIS

An indication of TDSB's commitment to GIS technology was its decision to implement an enterprise GIS at the administrative level. Prior to the merger of the seven school boards, GIS was being used sparingly by administrators in the planning, facilities management, and academic accountability departments.

After its formation, TDSB decided to incorporate GIS across the full range of its operations. Dr. Rob Brown of the Academic Accountability department recalls that "GIS technology was adopted because the complexity of the new TDSB necessitated new methods to examine and understand the organization and its component schools." The first GIS administrative effort resulted in the *Toronto District Schools Atlas.* The atlas is an interactive system centered around area maps for each of the elementary and secondary schools. These maps incorporate ancillary features of the school community such as libraries, community centers, day care centers, nursery schools, and private schools. The system also gives boundary data for the four education districts into which TDSB is divided and the thirty-one smaller educational units that are each led by a field superintendent. To date, nearly seven hundred maps have been developed, and several hundred printed. An Internet application of this system that uses ArcIMS technology is currently under development. When operational, the system will enable

superintendents to access the atlas from their desktops. Other projects conducted by the Academic Accountability team include school, student distribution, and food program maps. The department is also using GIS to better understand the impact of inner city indicators and to assist representative groups in studying boardwide amalgamation issues.

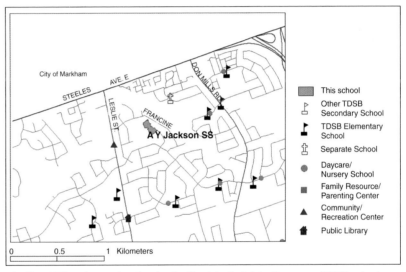

The first GIS project undertaken by the district administration was a TDSB school atlas. It was a familiar subject to cut their teeth on.

The Department of Planning and Facilities Management is another significant user of GIS. Since the amalgamation of the board, a powerful GIS platform has emerged. Specific projects include school bus routing; location analysis for unique programs, such as French immersion; and developing rationales for school boundaries and building closings.

Adopting GIS as a tool across all components of an educational system summoned up some unanticipated difficulties. Data acquisition proved to be especially problematical. With the advent of the District Site License and collaboration between educational and administrative sectors, negotiations began with outside data providers to acquire specific local data coverages.

The first negotiations occurred with the Planning Department of the new City of Toronto. An interagency agreement was reached whereby all new data acquisitions would be shared by departments, and most importantly from an educational perspective, be available for students in the classroom. This approach is cost effective and beneficial to a variety of stakeholders.

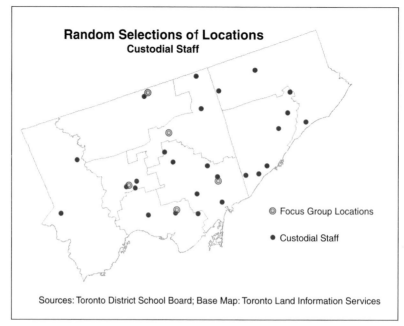

The Toronto District School Board manages public education for the fifth largest city in North America. The data is vast and the issues complex, making GIS a tool as necessary for administrators as it is important to students.

Classroom implications

Having a school district site license has had profound implications for students. Teachers are increasingly using the study of real-world problems as the centerpiece of their curriculum. Rather than being passive receivers of information, they see information—and consequently knowledge—as a two-way street. They can analyze, create, and disseminate findings. For example, when the Academic Accountability and Planning departments presented plans for the closing of schools and school boundary realignment, classes in the affected schools used GIS to evaluate and rank the different scenarios from a spatial perspective.

The Toronto District School Board, the City of Toronto, and the Province of Ontario have embarked on a unique collaboration between government and education. Never before have GIS and geotechnologies been adopted as the linchpin of an entire school board structure, nor served as a mandated core of the curriculum at this scale. The Toronto GIS experience will be an interesting story to follow as it unfolds chapter by chapter.

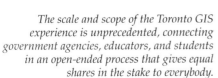

The scale and scope of the Toronto GIS experience is unprecedented, connecting government agencies, educators, and students in an open-ended process that gives equal shares in the stake to everybody.

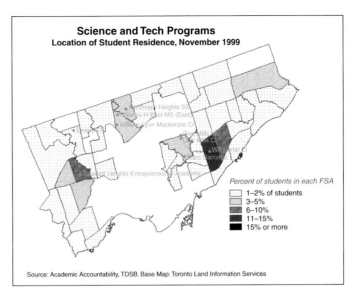

Science and Tech Programs
Location of Student Residence, November 1999

Percent of students in each FSA
- 1–2% of students
- 3–5%
- 6–10%
- 11–15%
- 15% or more

Source: Academic Accountability, TDSB. Base Map: Toronto Land Information Services

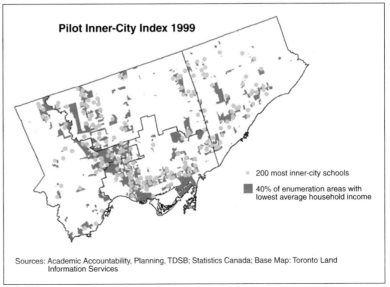

Pilot Inner-City Index 1999

- 200 most inner-city schools
- 40% of enumeration areas with lowest average household income

Sources: Academic Accountability, Planning, TDSB; Statistics Canada; Base Map: Toronto Land Information Services

Chapter 12

Connecting GIS and Problem Based Learning

Sarah Witham Bednarz, Assistant Professor
Texas A& M University
College Station, Texas

The use of GIS in the classroom calls for new approaches to curriculum-building, teaching, and the assessment of student performance. One highly effective method of exploring and thinking about the world with GIS (rather than simply learning how to manipulate the software) is to base all projects on a carefully chosen and complex problem. This method—Problem Based Learning (PBL)—encourages teachers and students to apply concepts and skills from other, nominally unrelated disciplines to investigate a real-world issue.

Working up problem scenarios that frame, focus, organize, and stimulate a nascent urge to learn, students conduct investigations while simultaneously learning research skills and the use of a variety of tools and technologies, including GIS. PBL–GIS closely mirrors the five skills of geography included in *Geography for Life: The National Geography Standards,* and addresses the approaches to inquiry recommended in the National Science Education Standards.

How does PBL–GIS work?

Classes developed according to the PBL–GIS model transform the classroom. Students take on new roles, directing their own work and "learning to learn" cooperatively as they investigate and solve problems. Students become project planners, collaborators, producers, and decision makers. They strengthen their understanding of ways to study geographic relationships, and learn about the practical side of making presentations. When you observe a PBL–GIS classroom you see motivated students who recognize the worth of their efforts. They enjoy taking on real-world issues and appreciate the autonomy that PBL–GIS fosters.

This doesn't mean, however, that the teacher is merely a spectator or cheerleader of student work. Some regard the teacher's many jobs in a PBL–GIS environment as even more complex than in a traditional classroom. The teacher designs the curriculum and supports student learning by serving as facilitator, coach, guide, and resource person who models inquiry. Instructors must have research methods and problem-solving strategies comfortably in hand, and possess reasonably proficient GIS skills, too.

GIS and Problem Based Learning is a way of cultivating different perspectives of time and space and the people moving through them.

Model one: The unstructured approach

A simple approach to PBL–GIS consists of four phases that should progress more or less seamlessly. (Students may have to revisit each stage's key questions more than once as they work through a problem.) The accompanying chart summarizes the stages and suggests how teachers can help students succeed.

Stage	Role of Teacher	Role of Student
Stage One Clarify topic. *What do we know?*	• Facilitator • Guide • Colearner with students	• Problem solver • Colearner *(discussing and sharing ideas)*

What is known?
Begin by giving students a topic related to the curriculum and connected to the standards. Encourage them to discuss the subject in a general way to clarify key terms, concepts, and fundamental issues. Find out with them what they already know about the topic. Organize those thoughts and revise prior understandings accordingly.

Stage	Role of Teacher	Role of Student
Stage Two Identify problem(s). *What needs to be known?*	• Guide *(to define problems)* • Resource *(to suggest data sources)* • Planning assistant *(working with students)* • Manager *(to assist in organizing student groups and work)*	• Planner • Decision maker • Problem solver

What needs to be known?
After an initial investigation and discussion, students must recast their topic as a problem, restating the key issues or concerns as questions, and generating preliminary ideas. Ask students to consider the information they need to investigate the question. What primary data should they collect? What secondary data is available? What level of analysis might be necessary to answer the question? Can GIS help to understand the issue?

"Think globally, act locally" can be turned around slightly to reflect the dynamic of PBL–GIS: "examine locally, think globally."

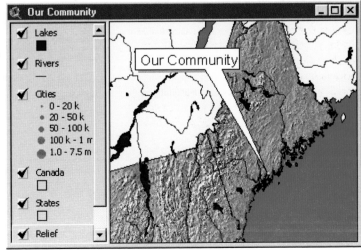

What is going to be done?
Students, usually working in small groups, will select the specific problem they want to explore, and decide on the resources they'll need and methods they'll use to complete their investigation. Ask them to describe the products they expect to generate as evidence of their problem-solving work, such as maps, reports, and Microsoft® PowerPoint® presentations. Help them to develop an action plan.

When students are choosing their problems or questions, be sure that they address five selection criteria. The problem or inquiry question must be: meaningful; important and central to their subject area; complex—that is, having no simple, obvious answer; relevant to the topic under study in class; and original.

It is important that students have clear plans and opportunities to articulate the concepts, skills, and methods they have learned. The typical product of PBL–GIS is a GIS-based map and an accompanying analysis and interpretation. Inferences from the maps help students to formulate generalizations, make decisions, and form judgments about the question, issue, or problem.

Stage	Role of Teacher	Role of Student

Stage Three
Plan inquiry.
What is going to be done?
	• Resource
	  (suggesting materials)
	• Model
	  *(GIS; problem solving;
	  research skills)*
	• Guide
	  *(check student work
	  plans; suggest strategies)*
	• Facilitator
	  *(coordinate resource use
	  and student interaction;
	  monitor progress;
	  anticipate resource
	  needs)*
		• Producer
		• Problem solver
		• Planner
		• Teacher
		  *(helping other students
		  with GIS and analysis)*

Meaningful, important, complex, relevant, original: when these elements come together in the context of community, paths of learning from one issue or subject to another become quite clear.

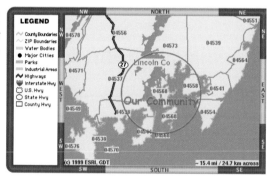

What has been learned?

Using PBL–GIS is one thing—real learning happens when students analyze their data and report results. In PBL–GIS, students are balancing three simultaneous tasks: they are investigating a topic and its associated problems, learning to use GIS, and making decisions about when and how to use GIS to conduct their inquiry. When students complete their research they prepare oral or written explanations of their findings and give reasons for their conclusions. Teachers will need to assist students in structuring their results and summary presentations.

Stage	Role of Teacher	Role of Student

Stage Four
Investigate and report.
What has been learned?
	• Model
	  (GIS; problem solving; research skills)
	• Guide
	  (manage presentations)
	• Facilitator
	  (check student findings)
		• Producer
		• Teacher
		  (sharing findings and teaching classmates; helping other students with technology)

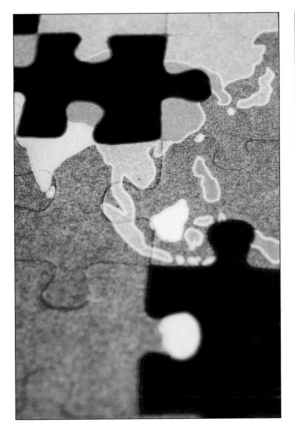

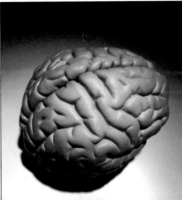

One of the most salient aspects of PBL–GIS work in the classroom is the way it encourages integration of facts and knowledge, not just simple accretion.

Model two: The structured approach

This model is a more tightly controlled version of PBL–GIS that give students more organization and less autonomy. The teacher has a more prominent classroom role. The process revolves around opportunities to investigate problems and issues that use available GIS data and basemaps. The CD–ROM that accompanies *GIS in Schools* contains examples that apply this approach.

Select a problem
See the chart at the right, "Selecting a topic," for guidelines on selecting a problem. Match the problem to national, state, and school district standards. Combine standards and curriculum objectives from more than one subject area such as geography, science, and language arts. Identify worthwhile, relevant, and useful problems that offer students numerous opportunities to work with significant content and acquire skills. Providing access to adequate resources that support in-depth investigations is an extremely important role of the teacher.

Selecting a topic

Selecting a good topic is crucial to the success of PBL–GIS. You want to use resources that you have but you also need to keep these criteria in mind as you refine your problem.

Scale
A good topic ranges from local to global concerns. For example, global climate change affects students' own communities as well as the world. Students can study issues at a range of scales. A topic that allows students to zoom in and out looking at data at a range of scales works well for PBL–GIS.

Relevancy
Good topics touch students' lives. Community-based research and service learning are motivating and compelling to most young people. They allow students to develop strong citizenship skills by working together to solve a problem. A topic that encourages students to generate their own problems and take action is good.

Continuity
An effective topic links the present to the past and gives students opportunities to consider the future. Looking at change over time is a powerful tool of analysis.

Ethics
A topic that features moral and ethical dimensions will give students opportunities to learn to use reason to make emotional decisions.

Interdisciplinary
A topic that allows the integration of several disciplines and skills is worth spending class time on. Make sure there are opportunities for a variety of tasks and that students have some choices. The focus should not be too narrow. Geography and environmental science topics lend themselves to cross-disciplinary exploration.

Complexity
A problem should be complex and open-ended enough to interest and motivate students but not too unwieldy and challenging.

Rigor
Select a topic that is rigorous, challenging, and interesting to students based on its complexity and issues, not just on the amount of information it encompasses.

Authenticity
Pick a topic that is rooted in the real world and which is significant because of its implications.

Develop a scenario

Place the problem in context by developing a stimulating rationale, event, or scenario that encourages student investigation. Provide some problem background in the scenario and include a description of why this work is important. Justify the investigation by linking it to the student's personal interests and concerns. Make the problem as directly related to the real world and exciting as possible. Specify the questions that students must define and answer during the course of their investigation. Let students know if they will work alone or in groups, and how the groups will be organized.

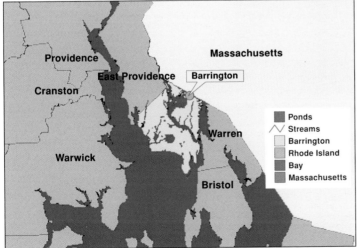

This work by an ambitious group of Barrington Middle Schoolers shows how focus can be sharpened as problem scenarios are discussed and developed.

Determine products and outcomes
Be clear about expectations for students. Decide what students will produce through their investigation and by what criteria these materials will be evaluated. Determine what tasks they must complete and if group products, individual work, or a combination of the two must be submitted. In addition to the requirement incorporating GIS, students may be asked to develop brochures, write memoranda, prepare proposals, illustrate research reports, construct exhibits, present their work to community members, or build Web sites that incorporate GIS-based findings.

Give clear information about how the assignments and student performance will be assessed. Again, check curriculum guidelines and standards to make sure students have appropriate opportunities to learn challenging content matter and skills. Consider available data resources and access to technology. Be clear and firm about placing students at the center of the problem solution.

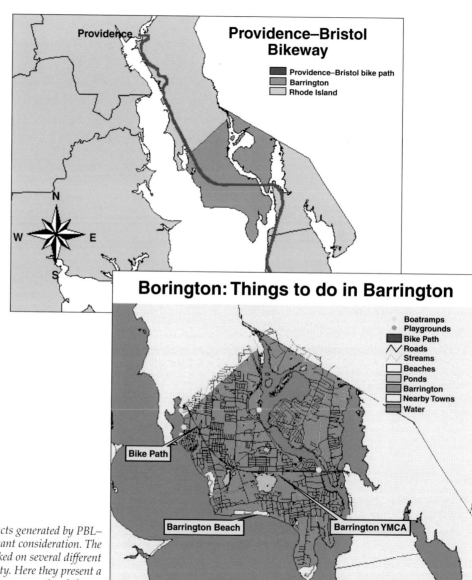

The usefulness of the products generated by PBL–GIS work is an important consideration. The Barrington group worked on several different levels of complexity. Here they present a recreational theme.

Develop a work plan

Outline how students will proceed with the investigation. Decide on what research strategies students will use. Use a calendar to schedule the progression of assignments and organize the activity flow. Make sure that students have adequate time to gain the competencies needed to address the problem. Guide, check, and monitor student progress through journals or other forms of self-reflection. Develop strategies to manage student access to resources and research tools they use in the investigation, including GIS. Balance modeling how to use GIS and think geographically with the opportunity for students to explore data and software on their own and learn from one another. Support students when they are frustrated but avoid doing the work for them.

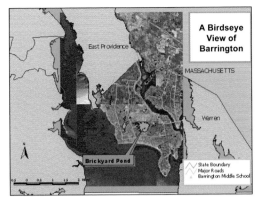

Here the Barrington group approaches the outer limits of geotechnology with satellite imagery. Spectacular as the images are, what's most impressive is the way the students were able to locate and use these extraordinary resources.

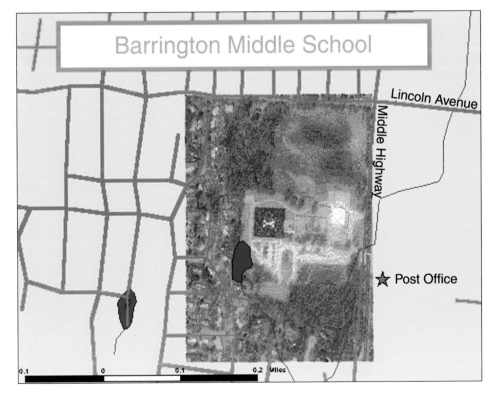

Communicate results and share and analyze products

Debrief and reflect with students when they have completed their projects. Are there alternative solutions to the problem? Are there different viewpoints on the issues? What content matter and skills did they learn? Did GIS help to answer questions? What are the students' assessment of their own performance? What learning strategies and responsibilities were most helpful?

This final stage of PBL–GIS may be organized as a culminating event in which students present their research findings to parents and community members. The last stage is very important. It cements student learning by making students acutely aware of their personal accomplishments.

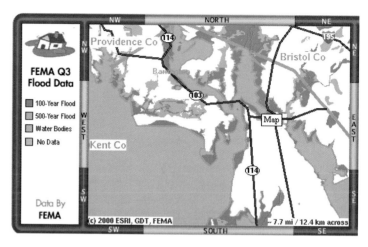

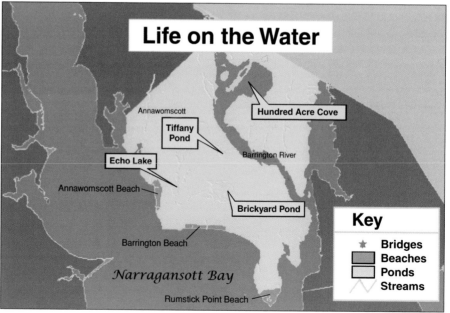

FEMA flood data added another level of complexity to the analytical work of the Barrington group, and provided another perspective of the community to be discussed at the end of the project.

Assessment in PBL–GIS

Evaluation in problem-based inquiry focuses on student's products and performance. Specific assessment criteria are presented during the early stages of a PBL–GIS activity to guide students. Scoring rubrics that clearly state expectations are helpful for students and send a clear message about standards. Be sure that the scoring guide measures all the outcomes that are valued, such as presenting a reasoned and sensible solution to the problem, communicating results, gaining content knowledge, and facility using GIS.

A comprehensive evaluation plan to manage a PBL–GIS investigation will incorporate a variety of assessment tools to ensure that students stay attentive and motivated. Daily grades can be given that are based on observations, student-group–teacher conferences, or journal entries.

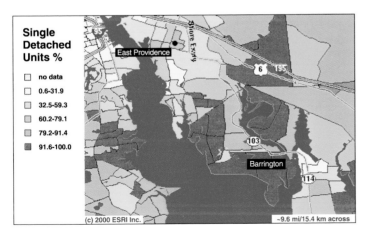

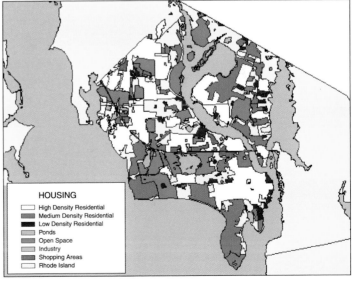

The Barrington Middle Schoolers' consideration of their community was rich, full, and varied—a natural outcome when tools are handled properly, interest is high, and guidelines are made clear.

Why PBL–GIS?

PBL–GIS is a bold but demonstrably effective way to structure the curriculum. It enhances learning, motivates and prepares students to make smooth school-to-work transitions, and makes teaching with GIS manageable.

When students learn with GIS, they apply three types of knowledge. They learn subject matter. This type of knowledge is called declarative knowledge. Declarative knowledge is "knowledge that . . ." or "knowledge about . . ." Secondly, students learn to use GIS software, for example, in displaying different layers of data or linking databases. Such skills are termed procedural knowledge and represent "knowledge how . . ." Thirdly, and, most importantly, students discover when and why to use certain procedures, such as queries, projections, and comparisons, to answer specific questions. This type of knowledge is called conditional knowledge, that is, knowing under what conditions, when, and why to apply a particular approach to solve problems. PBL–GIS offers students opportunities to master all three types of knowledge in authentic contexts.

Each case study in this book affirms how students benefit from working with GIS to solve real-world problems. Such work is exciting, engaging, and active, not boring, static, and passive. PBL–GIS prepares students for the 21st-century workplace. Skills valued by public- and private-sector employers are employed in PBL–GIS, including the abilities to work effectively with others, to apply problem-solving skills and inquiry strategies to real-world issues, and to effectively communicate through oral, written, and graphic expressions.

PBL–GIS connects students with their world and teaches them a marketable skill, spatial problem solving. It also teaches them a valuable perspective, how to think geographically. It's fun, too.

The Problem-Based Learning Rubric

	Understanding of problem	Reasoning to solve problem
4	• Demonstrates deep, coherent, and sophisticated understanding of the problem	• Analyzes the problem fully • Applies powerful problem-solving and thinking strategies • Uses information from a wide range of sources • Makes numerous and appropriate connections to individuals, events, and issues
3	• Demonstrates complete understanding of the problem	• Analyzes the problem • Applies appropriate problem-solving and thinking strategies • Uses information from a variety of sources • Makes appropriate connections to individuals, events, and issues
2	• Demonstrates partial understanding of the problem	• Partially analyzes the problem • Generally applies appropriate problem-solving and thinking strategies • Uses information from a limited number of sources • Attempts to make connections to individuals, events, and issues
1	• Misunderstands the problem	• Fails to organize or analyze the problem • Applies few and limited problem-solving and thinking strategies • Incorporates little supporting information • Makes no connections to individuals, events, and issues

References

American Association for the Advancement of Science. 1993. *Project 2061: Benchmarks for Scientific Literacy.* New York, New York: Oxford University Press.

Geography Education Standards Project. 1994. *Geography for Life: National Geography Standards.* Washington, D.C.: National Geographic Society.

Glasgow, N. A. 1997. *New Curriculum for New Times.* Thousand Oaks, California: Corwin Press.

Jones, B. F., C. M. Rasmussen, and M. C. Moffitt. 1997. *Real-Life Problem Solving.* Washington, D.C.: American Psychological Association.

National Research Council. 1996. *National Science Education Standards.* Washington, D.C.: National Academy Press.

Windschitl, M. 1999. "The Challenges of Sustaining a Constructivist Classroom Culture." Kappan 80 (10): 751–755.

Chapter 13

Student inquiry, GIS, and educational reform

*From the standpoint of the child, the great
waste in school comes from his inability
to utilize the experience he gets outside . . .
while on the other hand he is unable to apply
in his daily life what he is learning in school.
That is the isolation of school . . . its isolation
from life.*

— John Dewey

In recent years, the debate about the
impact and importance of using technology in the classroom has generally sub-
sided. Technology is alive and well in
schools all across America and computers
are now common educational tools.
Word-processing, spreadsheet, and
graphics programs have become standard applications that teachers and students use in their daily lives both in and
beyond the classroom.

An inherent aspect of technology is its tendency to evolve along new paths of interest. GIS, a relatively new blend of mapmaking and data management technologies, illustrates a new and powerful opportunity for schools. It creates spatial learning environments in which students can explore, analyze, and make decisions about problems in an interactive and challenging manner. The downside is that once again, teachers are being asked to work beyond the limits of their "computer comfort zone" and learn new software that requires considerable investments of time and energy. Why

should they do this? Can GIS help to revitalize and reform classroom practices?

The case studies in *GIS in Schools* have been selected to answer some of these questions. Each case study illustrates how the classroom curriculum can be enriched by using engaging activities that focus on communities. The teachers who developed these case studies have made considerable investments of their time and energy to create genuine learning experiences for their students. They have crafted GIS learning modules that redefine the roles of teacher and student. Traditional classrooms with activities led by

teachers do not appear among the case studies in this book. Rather, we see classrooms pictured as open spaces in which teaching and learning swirl and flow among students and their teachers.

Educational researchers have discovered that students learn best (i.e., are able to transfer their understanding) through directed inquiry, collaboration among peers, use of tools, and in rich contextualized settings. These elements of ideal learning environments appear repeatedly in the *GIS in Schools* case studies.

The introduction of new technologies and new software can decrease the "computer comfort zone" in the classroom, but studies indicate the problem is greater for teachers than it is for students—even the youngest—who can seem almost nonchalant in their familiarity with these new tools.

Learn with or about GIS?

During the past few years, debate over GIS and education has shifted between whether students should learn with or about GIS.

• Learning about GIS is the more common approach found in classrooms today. It focuses on the technological skills of GIS such as data handling and information management with little emphasis on spatial problem-solving methodology.

• Learning with GIS emphasizes the process of spatial inquiry and learning to reason spatially. The focus is on using GIS as a tool for spatial problem solving. This instructional model integrates the questions and tasks central to geographic understanding that are documented in *Geography for Life: National Geography Standards 1994*.

What the stories in this book clearly show is that teachers are making decisions on a case-by-case basis about how GIS can be most appropriately applied. The essential question is "does the issue have spatial characteristics?" If there is a spatial context, then GIS may be a tool that helps students to learn or solve problems. The needs of the student and the connections to the curriculum are the twin forces driving the implementation of GIS in classrooms.

The GIS lab can be a fun place to be—almost a game room—but the emphasis must always be clearly placed on problem solving.

But, as GIS becomes increasingly common as a general tool for teaching and learning, questions will eventually rise as to whether GIS is helping to achieve education reform. Under the banner of education reform we include all of the major recommendations for changes in content and teaching that are spelled out in the 1990s curriculum reform documents, such as *Science For All Americans, Math Counts!, Project 2061: Benchmarks for Science Literacy,* and *Geography for Life.*

What is novel about the current wave of reform is that it is systemic. It rests on the premise that unless everything changes, nothing changes. Change originates in policy decisions such as those that involve national curriculum standards. These policy statements have resulted in innovative educational programs that incorporate new approaches to teaching and learning. What gives today's reform initiatives a higher probability of success are new understandings about how people learn, an educational system that is increasingly connected with standards, adoption of approaches to teaching and learning based in inquiry, and the widespread classroom use of technology. The *GIS in Schools* case studies provide convincing evidence that GIS complements each of these pieces of the education reform puzzle.

The role that GIS plays in education reform, and in compliance with new standards, depends on the relation of the technology to "current events." Once timeliness and relevance are established, paths to other areas of learning become clearer and more accessible.

National education standards

Every teacher whose story is told in these case studies shares a love of working with students and an innovative teaching style. Since benchmarks have been developed for most subject areas, standards have become the cornerstone of every teacher's professional life.

Standards are an outgrowth of the landmark 1983 publication, *A Nation at Risk.* This report painted a bleak picture of the state of American education and urged the country to unite in reforming its educational system. Standards have rigor, address both curriculum content and approaches to instruction, and are closely tied to student assessment. Standards are not the same as a curriculum nor do they prescribe how to teach. Standards tell us what students need to know and by what point they need to know it.

Standards are in essence maps to "literacy," whether that literacy is in areas of mathematics, the social sciences, language arts, or science. In science and geography, student-centered inquiry is expected to serve as a central approach to the learning experience. Knowledge about the environment and how human decisions affect our world are major pieces of the science standards. Geography standards state that geographically informed people should know how to use tools and technologies to "acquire, process, and report information from a spatial perspective." The fit between GIS applications and the standards in these content areas is very close. The *GIS in Schools* case studies provide vivid illustrations of what students are doing in GIS-enhanced classrooms and how this technology is helping to meet the goals of education reform.

The spatial perspective that geography standards speak of can be translated to mean (among other things) a broadened and well-informed world view.

How people learn

Most contemporary ideas about learning originated during the past twenty-five years. Today, there is a science of learning that draws from cognitive psychology, brain research, and developmental science. We know that student learning is strongly influenced by prior knowledge, affected by peer interaction, and enhanced by actively solving problems that the learner finds interesting and relevant. (The criteria for judging interest and relevancy remain problematic, however.)

The goal of learning today is for students to gain usable knowledge, such as problem-solving skills, and develop the ability to apply these skills in a variety of situations. There is considerable evidence of each of these new perspectives in the GIS case studies. The educational image we see of the new student is that of an active worker, the person who ultimately makes sense of an educational experience. The role of the new teacher is coach, facilitator, guide, or mentor. The new curriculum is sensitive to individual learning styles, built around student inquiry, integrates subject areas, and emphasizes learning in collaborative groups.

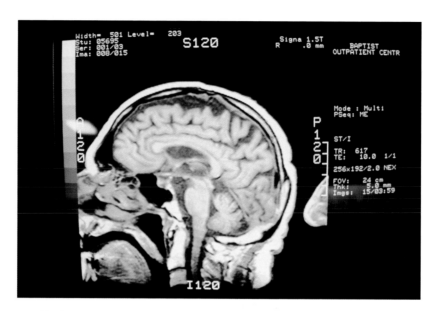

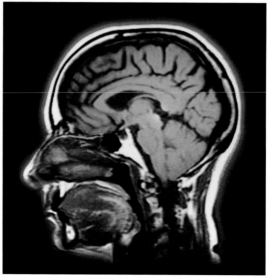

The elements and processes of human learning are complex and only just beginning to be understood. Desire to learn, however, and what promotes that desire is perhaps the key to knowledge.

A final word

A classroom that uses GIS as a problem-solving tool is a classroom in which the walls are invisible and the teacher and student assume roles that are nontraditional. Textbooks and teacher-centered activities are replaced by "pods" of students engaged in different projects. Technological tools, digital data, Internet access, paper maps, community resource people, and noisy conversations fill the classroom. Adopting this technology is not for the fainthearted. But, integrating GIS into the curriculum rewards teachers by creating intellectually challenging and demanding learning opportunities. We hope that *GIS in Schools* presents images that are active, exciting, and centered on learning.

GIScience

GIS for Everyone
Now everyone can create smart maps for school, work, home, or community action using a personal computer. Includes the ArcExplorer™ geographic data viewer and more than 500 megabytes of geographic data. ISBN 1-879102-49-8

The ESRI Guide to GIS Analysis
An important new book about how to do real analysis with a geographic information system. *The ESRI Guide to GIS Analysis, Volume 1: Geographic Patterns and Relationships* focuses on six of the most common geographic analysis tasks. ISBN 1-879102-06-4

Modeling Our World
With this comprehensive guide and reference to GIS data modeling and to the new geodatabase model introduced with ArcInfo™ 8, you'll learn how to make the right decisions about modeling data, from database design and data capture to spatial analysis and visual presentation. ISBN 1-879102-62-5

Hydrologic and Hydraulic Modeling Support with Geographic Information Systems
This book presents the invited papers in water resources at the 1999 ESRI International User Conference. Covering practical issues related to hydrologic and hydraulic water quantity modeling support using GIS, the concepts and techniques apply to any hydrologic and hydraulic model requiring spatial data or spatial visualization. ISBN 1-879102-80-3

Beyond Maps: GIS and Decision Making in Local Government
Beyond Maps shows how local governments are making geographic information systems true management tools. Packed with real-life examples, it explores innovative ways to use GIS to improve local government operations. ISBN 1-879102-79-X

ESRI Map Book: Applications of Geographic Information Systems
A full-color collection of some of the finest maps produced using GIS software. Published annually since 1984, this unique book celebrates the mapping achievements of GIS professionals. ISBN 1-879102-60-9

The Case Studies Series

ArcView GIS Means Business
Written for business professionals, this book is a behind-the-scenes look at how some of America's most successful companies have used desktop GIS technology. The book is loaded with full-color illustrations and comes with a trial copy of ArcView GIS software and a GIS tutorial. ISBN 1-879102-51-X

Zeroing In: Geographic Information Systems at Work in the Community
In twelve "tales from the digital map age," this book shows how people use GIS in their daily jobs. An accessible and engaging introduction to GIS for anyone who deals with geographic information. ISBN 1-879102-50-1

Serving Maps on the Internet
Take an insider's look at how today's forward-thinking organizations distribute map-based information via the Internet. Case studies cover a range of applications for ArcView Internet Map Server technology from ESRI. This book should interest anyone who wants to publish geospatial data on the World Wide Web. ISBN 1-879102-52-8

Managing Natural Resources with GIS
Find out how GIS technology helps people design solutions to such pressing challenges as wildfires, urban blight, air and water degradation, species endangerment, disaster mitigation, coastline erosion, and public education. The experiences of public and private organizations provide real-world examples. ISBN 1-879102-53-6

Enterprise GIS for Energy Companies
A volume of case studies showing how electric and gas utilities use geographic information systems to manage their facilities more cost effectively, find new market opportunities, and better serve their customers. ISBN 1-879102-48-X

More ESRI Press titles are listed on the next page ➤

ESRI Press
380 New York Street
Redlands, California 92373-8100

www.esri.com/esripress

ESRI educational products cover topics related to geographic information science, GIS applications, and ESRI technology. You can choose among instructor-led courses, Web-based courses, and self-study workbooks to find education solutions that fit your learning style and pocketbook. Visit www.esri.com/education for more information.

Other books from ESRI Press

The Case Studies Series CONTINUED

Transportation GIS
From monitoring rail systems and airplane noise levels, to making bus routes more efficient and improving roads, this book describes how geographic information systems have emerged as the tool of choice for transportation planners. ISBN 1-879102-47-1

GIS for Landscape Architects
From Karen Hanna, noted landscape architect and GIS pioneer, comes *GIS for Landscape Architects*. Through actual examples, you'll learn how landscape architects, land planners, and designers now rely on GIS to create visual frameworks within which spatial data and information are gathered, interpreted, manipulated, and shared. ISBN 1-879102-64-1

GIS for Health Organizations
Health management is a rapidly developing field, where even slight shifts in policy affect the health care we receive. In this book, you'll see how physicians, public health officials, insurance providers, hospitals, epidemiologists, researchers, and HMO executives use GIS to focus resources to meet the needs of those in their care. ISBN 1-879102-65-X

GIS in Public Policy
This book shows how policy makers and others on the front lines of public service are putting GIS to work—to carry out the will of voters and legislators, and to inform and influence their decisions. *GIS in Public Policy* shows vividly the very real benefits of this new digital tool for anyone with an interest in, or influence over, the ways our institutions shape our lives. ISBN 1-879102-66-8

Integrating GIS and the Global Positioning System
The Global Positioning System is an explosively growing technology. *Integrating GIS and the Global Positioning System* covers the basics of GPS technology and presents several case studies that illustrate some of the ways in which the power of GPS is being harnessed to the depth of GIS: accuracy in measurement and completeness of coverage. ISBN 1-879102-81-1

ESRI Software Workbooks

Understanding GIS: The ARC/INFO® Method (UNIX®/Windows NT® version)
A hands-on introduction to geographic information system technology. Designed primarily for beginners, this classic text guides readers through a complete GIS project in ten easy-to-follow lessons. ISBN 1-879102-01-3

Understanding GIS: The ARC/INFO Method (PC version)
ISBN 1-879102-00-5

ARC Macro Language: Developing ARC/INFO Menus and Macros with AML
ARC Macro Language (AML™) software gives you the power to tailor workstation ARC/INFO software's geoprocessing operations to specific applications. This workbook teaches AML in the context of accomplishing practical workstation ARC/INFO tasks, and presents both basic and advanced techniques. ISBN 1-879102-18-8

Getting to Know ArcView GIS
A colorful, nontechnical introduction to GIS technology and ArcView GIS software, this workbook comes with a working ArcView GIS demonstration copy. Follow the book's scenario-based exercises or work through them using the CD and learn how to do your own ArcView GIS project. ISBN 1-879102-46-3

Extending ArcView GIS
This sequel to the award-winning *Getting to Know ArcView GIS* is written for those who understand basic GIS concepts and are ready to extend the analytical power of the core ArcView GIS software. The book consists of short conceptual overviews followed by detailed exercises framed in the context of real problems. ISBN 1-879102-05-6

ESRI Press
380 New York Street
Redlands, California 92373-8100
www.esri.com/esripress

ESRI Press publishes a growing list of GIS-related books. Ask for these books at your local bookstore or order by calling 1-800-447-9778. You can also shop online at www.esri.com/gisstore. Outside the United States, contact your local ESRI distributor.